Signals and Communication Technology

This series is devoted to fundamentals and applications of modern methods of signal processing and cutting-edge communication technologies. The main topics are information and signal theory, acoustical signal processing, image processing and multimedia systems, mobile and wireless communications, and computer and communication networks. Volumes in the series address researchers in academia and industrial R&D departments. The series is application-oriented. The level of presentation of each individual volume, however, depends on the subject and can range from practical to scientific.

Indexing: All books in "Signals and Communication Technology" are indexed by Scopus and zbMATH

For general information about this book series, comments or suggestions, please contact Mary James at mary.james@springer.com or Ramesh Nath Premnath at ramesh.premnath@springer.com.

Lei Meng
Editor

3rd International Conference on Cloud Computing and Computer Networks

CCCN 2025

Springer

Editor
Lei Meng
School of Software
Shangdong University
Jinan, China

ISSN 1860-4862 ISSN 1860-4870 (electronic)
Signals and Communication Technology
ISBN 978-3-032-03631-5 ISBN 978-3-032-03632-2 (eBook)
https://doi.org/10.1007/978-3-032-03632-2

This Springer imprint is published by the registered company Springer Nature Switzerland AG
The registered company address is: Gewerbestrasse 11, 6330 Cham, Switzerland

Preface

The 3rd International Conference on Cloud Computing and Computer Networks (CCCN 2025) was held in Singapore from 16 to 18 May 2025, and we are pleased to present to you the proceedings of this conference. This conference is a melting pot of ideas and also promotes in-depth discussions and knowledge exchange in the field of Cloud Computing and Computer Networks.

Cloud computing and computer networks have become the cornerstone of modern information technology infrastructure. Cloud computing provides scalable, on-demand computing resources and services, and computer networks provide critical connectivity that facilitates seamless communication and data transfer on a global scale. The rapid development of these technologies has spawned a plethora of new applications and services, ranging from big data analytics and artificial intelligence to the Internet of Things and edge computing.

The papers included in this volume of proceedings cover a wide spectrum of topics, reflecting the latest research trends and advancements in cloud computing and computer networks. From novel algorithms and architectures for cloud resource management and network optimization to security and privacy issues in cloud environments, from the integration of cloud computing with emerging technologies to the exploration of future directions for network development, each paper presents unique insights and innovative solutions. These contributions not only demonstrate the high academic and technical level of the research in these fields but also offer valuable references for further study and practical applications.

We would like to express our sincere gratitude to all the authors who have submitted their work to the conference. Their dedication and hard work have significantly enriched the content of this proceedings. We also extend our heartfelt thanks to the members of the program committee, whose meticulous review and professional advice have ensured the quality and relevance of the selected papers.

We believe this volume of proceedings will serve as an important resource for researchers, students, and professionals in the fields of cloud computing and computer networks. It will inspire further research, promote technological innovation, and contribute to the development of more efficient, secure, and intelligent cloud computing and computer network systems. As we look to the future, we are confident that the continuous exploration and innovation in these fields will bring about even more remarkable breakthroughs and play an increasingly important role in shaping the digital world.

We hope that readers will find the papers in this volume of proceedings both informative and inspiring, and we look forward to your continued support and participation in the future.

Lei Meng
Proceedings Editor

Conference Committees

General Chairs

Lei Meng	Shangdong University, China
Jianqiang Li	Shenzhen University, China

Conference Co-chair

Xiaoli Li	Institute for Infocomm Research (I2R), A*STAR, Singapore

Program Chairs

Hong Shen	Sun Yat-sen University, China
Tok Wang Ling	National University of Singapore, Singapore
Huiru (Jane) Zheng	Ulster University, UK
Liying Zheng	Harbin Engineering University, China

Publicity Chairs

Hui Liu	Central South University, China
Somkiat Tangjitsitcharoen	Chulalongkorn University, Thailand

Local Chairs

Teo Tee Hui	Singapore University of Technology and Design, Singapore
Kok Chiang Liang	Newcastle Australia Institute of Higher Education, Singapore

Local Co-chair

Koh Yit Yan	Newcastle Australia Institute of Higher Education, Singapore

Technical Committees

Adriano Péres	Universidade Federal de Santa Catarina - UFSC Blumenau, UK
B. V. A. N. S. S Prabhakar Rao	Vellore Institute of Technology, India
Bernd E. Wolfinger	University of Hamburg, Germany
Casey How	Singapore University of Social Sciences, Singapore
Chen Wang	Huazhong University of Science and Technology, China
Chien-Chang Chou	National Kaohsiung University of Science and Technology, Taiwan
Chin Cheng Siong	Newcastle University in Singapore, Singapore
Francesco Zirilli	Sapienza Universita Roma, Italy
Dr. JE. Chukwuere	North West University, South Africa
Jianbiao Zhang	Beijing University of Technology, China
Jun Zhao	Nanyang Technological University
June Tay	Singapore University of Social Sciences, Singapore
Jingzheng Ren	Hong Kong Polytechnic University, China
Krzysztof Ejsmont	Warsaw University of Technology, Poland
Lakmali Shashika Karunarathne	York St John University, London campus, UK
Ljiljana Trajkovic	Simon Fraser University, Canada
Liu Ziwen	Singapore University of Social Sciences, Singapore
Maki Habib	The American University in Cairo, Egypt
Patrizio Dazzi	Information Science and Technologies Institute (ISTI), Italy
Qiang Duan	Pennsylvania State University, USA
Ran-Dang Hoa	Kumoh National Institute of Technology, South of Korea
Seppo Sirkemaa	University of Turku, Pori Unit, Finland
Venkata Ramarao Sanka	Johnson & Johnson, USA
Walaa H. Elashmawi	Misr International University, Egypt
Wang Gang	Hefei University of Technology, China
Yewkee Wong	Hong Kong Chu Hai College, Hong Kong, China

Yilun Shang	Northumbria University, UK
Yousuf Nasser Al Husaini	Arab Open University, Oman
Zhu Yongqing	Singapore University of Social Sciences, Singapore

Contents

Computer Network and Information Security

Next Generation IoT and Communication Network

Intelligent Image Analysis and Processing Method

Lung Cancer Detection and Classification with Deep Learning

Chiang Liang Kok(✉) and Cheng Yao Song

The University of Newcastle, Newcastle, Australia
{chiangliang.kok,chengyao.song}@newcastle.edu.au

Abstract. Lung cancer remains a leading cause of cancer-related deaths, emphasizing the need for early and accurate diagnosis. Advances in AI-driven medical imaging, particularly CT scans, have significantly improved detection and classification while reducing the workload of radiologists. By automating repetitive tasks and prioritizing high-risk cases, deep learning systems enable faster, more consistent analyses, allowing clinicians to focus on complex decision-making.

For lung cancer detection, traditional machine learning methods like Support Vector Machines (SVM) and K-Nearest Neighbor (KNN) have pros and cons [13], while deep learning models such as CNNs demonstrate superior performance, reaching 94.11% accuracy. However, performance varies with dataset size and model tuning, as seen in a CNN with 86.06% accuracy when trained on larger datasets. In lung cancer classification, deep learning models excel in distinguishing between benign, malignant, and subtypes like Adenocarcinoma (AC) and Squamous Cell Carcinoma (SCC). ResNet50 achieves the highest accuracy (98%), followed by VGG19 (97%) and EfficientNetB7 (96%). Traditional methods like the Johnson Reducer (JR) and Repeated Incremental Pruning to Produce Error Reduction (RIPPER) algorithms also perform well, with 95% accuracy for SCC classification. Comparative studies confirm that CT scans outperform PET scans in classification tasks, reinforcing their diagnostic superiority. These findings highlight the transformative potential of AI in lung cancer diagnostics, not only in improving accuracy but also in streamlining clinical workflows. However, the importance of model optimization and integration into real-world practice remains critical.

Keywords: deep learning · CT scan · Pet scan · Detection · Classification

1 Introduction

Lung cancer is one of the deadliest cancers worldwide, where early detection and precise classification drastically improve survival rates. Medical imaging, particularly CT scans, plays a pivotal role in identifying suspicious nodules [8, 13], while AI techniques enhance diagnostic accuracy through automated analysis. For detection AI-based lung cancer detection employs both traditional machine learning such as support vector machine(SVM) and K-nearest neighbor (KNN) have their strength and weakness however by combining them with deep learning models it can significantly improve their

L. Meng (Ed.): CCCN 2025, SCT, pp. 3–10, 2025.
https://doi.org/10.1007/978-3-032-03632-2_1

accuracy. For example by using GoogLeNet with SVM and different kernels achieves up to 92.12% accuracy, while GoogLeNet with KNN reaches 92.45% [1]. However, other deep learning models, particularly CNNs, outperform these methods, with a 6-layer CNN achieving 94.11% accuracy—surpassing ANN (90.24%) and MLP (92.12%). Performance depends on dataset size and tuning, as evidenced by another CNN model with 86.06% accuracy despite a larger training set, suggesting potential underfitting. Classification after detection, AI enables precise classification of lung nodules into normal, benign, or malignant categories. Deep learning models like ResNet50 (98% accuracy), VGG19 (97%), and EfficientNetB7 (96%) demonstrate remarkable performance. Traditional methods, such as JR and RIPPER algorithms, also achieve high accuracy (95% for SCC). Notably, CT scans consistently outperform PET scans in classification tasks, with one study reporting 92.3% vs. 83.1% accuracy for the same CNN model. These advancements underscore AI's transformative role in lung cancer diagnostics. However, variations in model performance highlight the need for careful architecture selection, sufficient training data, and hyperparameter optimization to maximize accuracy in both detection and classification tasks.

2 Literature Review

Lung cancer diagnosis have improved significantly however it still mainly focuses on this 2 primary task: detection and classification.

Detection

Early and accurate detection of lung cancer is critical for improving patient outcomes and integrating medical imaging such as CT scans. As CT scans can provide high-resolution 3D images by combining multiple layer of 2D images which improves the accuracy of the identification of small nodules that may indicate early-stage cancer. Which helps in detecting subtle patterns and anomalies [1, 3, 7, 17].

Classification

Unlike detection, which identifies the presence of abnormalities in CT scans, classification extracts detailed information from the images to categorize lung nodules into distinct groups, such as normal, benign, or malignant. Advanced machine learning and deep learning algorithms analyze features like size, shape, texture, and growth patterns to differentiate between non-cancerous lesions and aggressive tumors. By using Adeno Carcinoma (AC) and Squamous cell Cancer (SCC) samples then enhance the image and extract features, then it input Gene expression profile to compare and use Johnson Reducer (JR) Algorithm and Repeated Incremental Pruning to Produce Error Reduction (RIPPER) [5, 18].

We evaluate and compares the performance of different types of AI such as Artificial Neural Network (ANN) Convolutional Neural Networks (CNNs), GoogLeNet and ResNet50. Additionally, we explore hybrid approaches that integrate these models with traditional methods such as SVM and KNN with image enhancement system such as Contrasted Limited Adaptive Histogram Equalization (CLAHE) to further enhance its accuracy [11]. Most of the dataset used was from LUng Nodule Analysis (LUNA) and The Cancer Imaging Archive [12].

3 Performance Evaluation of Different AI

3.1 Detection

By combining GoogLeNet with support vector machine (SVM) or K-Nearest Neighbors (KNN) to detect and extract abnormal features such as cancer and filter the result to improve the accuracy with different type of kernel. SVM with linear have a 91.45% accuracy and 0.9926 F-score, SVM with cubic have a 92.12% accuracy 0.9843 F-score, SVM with Quadratic have a 89.77% accuracy and 0.9843 F-score. KNN with Weightage have 92.41% accuracy and 0.9689 F-score, KNN with cubic have a 91.34% accuracy and 0.9505 F-score, KNN with medium have a 92.45% accuracy and 0.9451 F-score [1]. Similar to the one without AI KNN with weighted have the highest accuracy but SVM with cubic have higher accuracy than linear from their study [13] (Fig. 1).

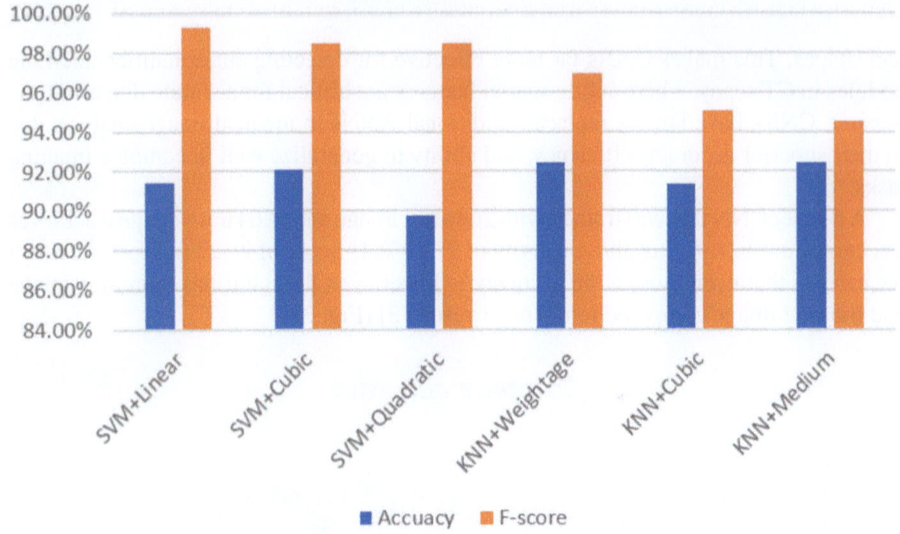

Fig. 1. GoogLeNet with SVM vs GoogLeNet with KNN

Other deep learning system a total of 343 images are used for training while 112 images are used for testing with the result of a 6 layer CNN it achieves the result of 94.11% accuracy and by comparing it with ANN of a 90.24% accuracy and Multilayer Perceptron (MLP) of a 92.12% accuracy from [2, 15] (Fig. 2).

From the result we can see that CNN have a higher accuracy than MLP and ANN as ANN and MLP, are capable of learning complex patterns from data, they are not well-suited for image-based tasks like medical imaging. ANNs treat images as flat input vectors, ignoring the spatial structure and local pixel relationships that are essential for recognizing visual patterns. This leads to many parameters, increased computational cost, and poor performance on high-dimensional image data. In contrast, CNN are specifically designed to handle image inputs by using convolutional layers that preserve spatial relationships and automatically learn important visual features such as edges, textures,

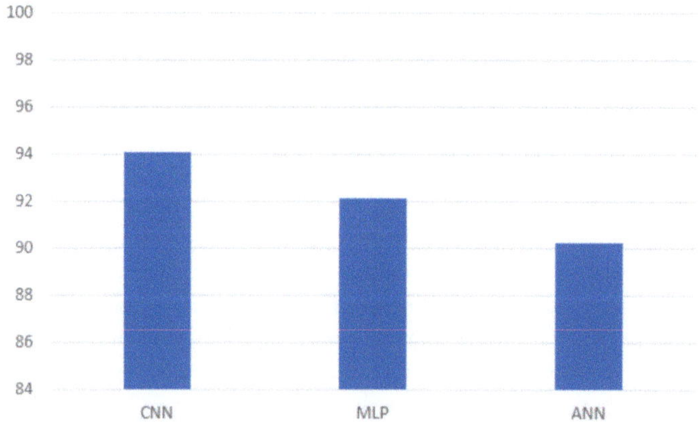

Fig. 2. Compare different type of deep learning system

and shapes. This makes CNNs far more effective for detecting abnormalities like lung nodules in CT scans, where subtle spatial features are critical for accurate diagnosis. As a result, CNNs have largely replaced traditional ANNs in medical image analysis due to their superior accuracy, efficiency, and ability to generalize well in complex imaging tasks.

A 6-layer CNN that was trained with 2686 test image and 3864 train image it achieves a accuracy of 86.06%, Recall/Sensitivity of 97.14%, F-Score of 91.27%. From the result we can see that even though the accuracy is lower even though it has more test image and training image compared to the other result [3] (Fig. 3).

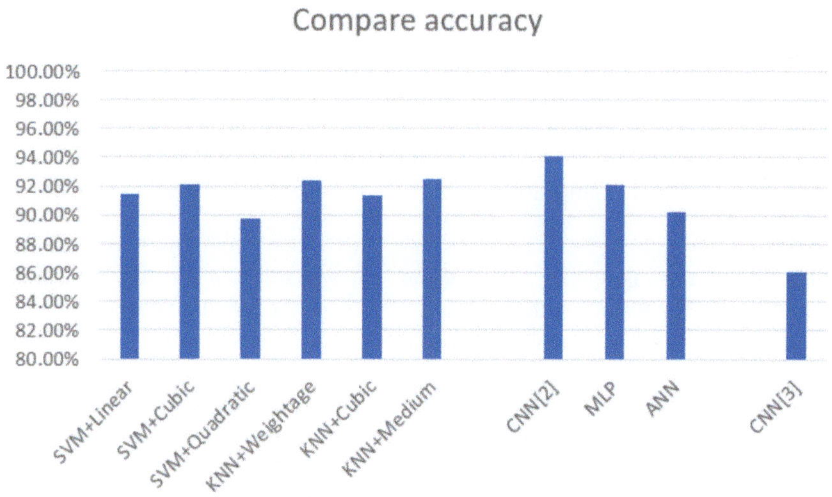

Fig. 3. Comparing accuracy

By comparing SVM, KNN with deep learning system we can see that deep learning system from [2] have better accuracy than SVM and KNN from [1] however the CNN from [3] have lower accuracy. Which can be caused by underfitting, and it requires more fine tuning.

3.2 Classification

CT scan uses multiple 2D layer image and stack them to show a 3D image that provides a better visual aid. However, PET scan uses tracer injection which during the scan the cancer cell will aggressively absorb more tracer which will display as a brighter color during the scan.

Using different type of AI such as AlexNet14, CNN-T5 3D and LeNet. CNN 3D is a more advance type of CNN, normal CNN in imaging uses height and width 3D incudes depth which help to improve the accuracy [10, 19, 20]. They uses Adeno Carcinoma (AC) and Squamous cell Cancer (SCC) samples then enhance the image and extract features, then it input Gene expression profile to compare and use Johnson Reducer (JR) Algorithm and Repeated Incremental Pruning to Produce Error Reduction (RIPPER) Algorithm to filter and classification. The Lung AC sample achieved a accuracy of 92% and Lung SCC sample achieved a accuracy of 95% [4]. Like previous result, however they test AC and SCC which was obtained from CT scan and PET scan and uses the same CNN to compare the result. Which CT scan has a accuracy of 92.3% and 83.1% which shows that CT scan has better accuracy with classification [5] (Figs. 4 and 5).

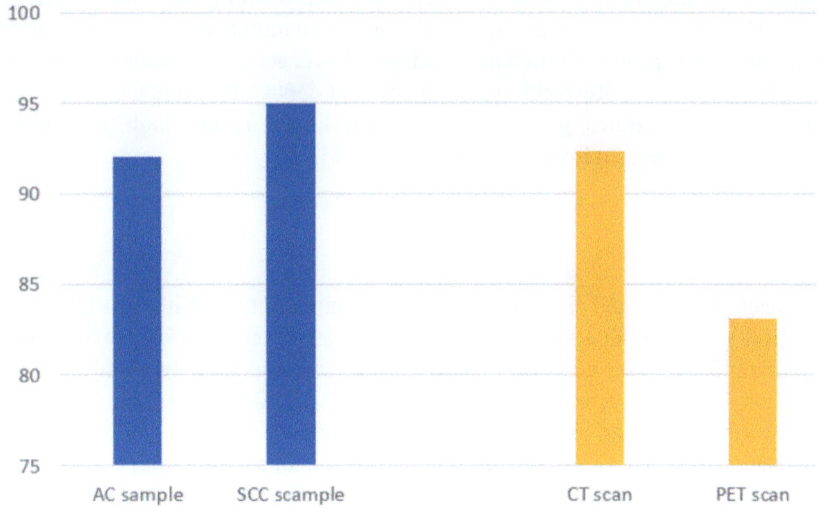

Fig. 4. Comparing CT scan and PET scan

ResNet50 achieves the highest accuracy primarily due to its residual connections, which allow the network to train very deep layers without suffering from the vanishing gradient problem. These skip connections help the model learn more complex features, improving its ability to detect subtle patterns in medical images like CT scans. VGG19, while also deep, lacks residual connections and uses many parameters with simple stacked convolution layers, making it less efficient and slightly less accurate despite

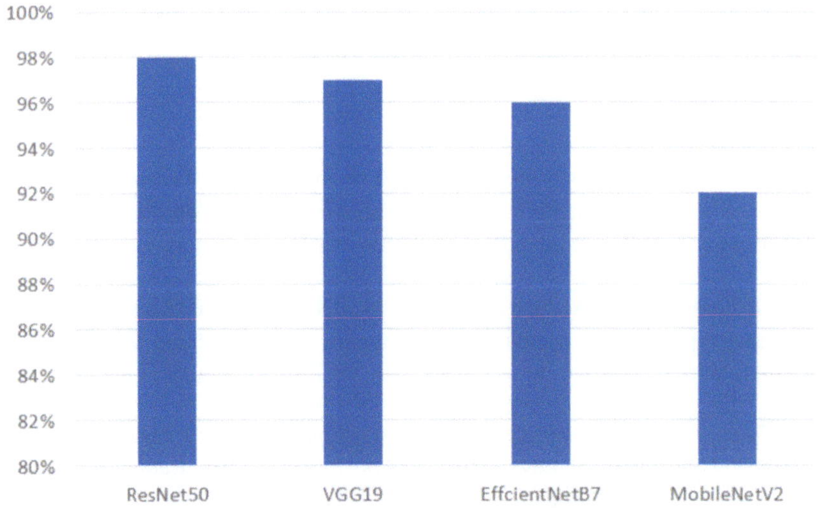

Fig. 5. Deep learning system accuracy

its depth. EfficientNetB7 is highly optimized for parameter efficiency and scales depth, width, and resolution systematically. Although it's very powerful, it may not outperform ResNet50 if the dataset or task favors deeper residual learning over model efficiency. MobileNetV2, designed for lightweight performance on mobile devices, trades off depth and complexity for speed and efficiency, leading to lower accuracy compared to the heavier models. In summary, ResNet50's combination of depth and residual learning allows it to capture richer features, giving it an edge in tasks that require high sensitivity and precision, such as medical image classification [7, 8].

4 Conclusion

Lung cancer detection and classification are both important and with the old algorithm method still provide high accuracy of 92.12% for detection and 95% for SCC sample for classification, with the addition of deep learning some neural network like ResNet50 have achieved a higher accuracy of 98% and CNN of 94.11% for detection. However, we can also see that the neural network can also lead to lower accuracy if not fine-tuned. We also see that based on the accuracy of CT scan vs Pet scan with the same amount of data and neural network CT scan have a higher accuracy which proves that it can provides better aid to radiology. However, there will still be some error which the radiology need to manually input the correct result to further improve the overall accuracy, with the dataset increases AI layer will also increase to provide better accuracy which means better hardware are required to support it.

References

1. Shafiq, S., Asghar, M.A., Amjad, M.E., Ibrahim, J.: An effective early stage detection of lung cancer using fuzzy local information cMean and GoogLeNet. In: 2022 16th International Conference on Open Source Systems and Technologies (ICOSST), Lahore, Pakistan, pp. 1–6 (2022). https://doi.org/10.1109/ICOSST57195.2022.10016866
2. Aharonu, M., Kumar, R.L.: Convolutional neural network based framework for automatic lung cancer detection from lung CT images. In: 2022 International Conference on Smart Generation Computing, Communication and Networking (SMART GENCON), Bangalore, India, pp. 1–7 (2022). https://doi.org/10.1109/SMARTGENCON56628.2022.10084235
3. Saravanan, T.M., Jagadeesan, M., Selvaraj, P.A., Mubarak, M.M., Kumar, S.P., Sanjay, S.: A CNN based machine learning scheme to detect lung cancer detection from CT scan images. In: 2023 International Conference on Computer Communication and Informatics (ICCCI), Coimbatore, India, pp. 1–5 (2023). https://doi.org/10.1109/ICCCI56745.2023.10128479
4. Vikruthi, S., Thippagudisa, K.B., Basha, P.H., Vamsi, P., Chandra, G.M., Kiran, V.S.: Performance analysis of gene expression profiles of lung cancer prediction using JR algorithm. In: 2023 International Conference on Sustainable Communication Networks and Application (ICSCNA), Theni, India, pp. 1731–1736 (2023). https://doi.org/10.1109/ICSCNA58489.2023.10370661
5. Suma, K.V., Sonali, C.S., Chinmayi, B.S., Easa, M.: CNN models comparison for lung cancer classification using CT and PET scans. In: 2022 IEEE 2nd Mysore Sub Section International Conference (MysuruCon), Mysuru, India, pp. 1–5 (2022). https://doi.org/10.1109/MysuruCon55714.2022.9972704
6. Khatun, M.F., Ajmain, M.R., Assaduzzaman, M.: A deep learning approach to detect and classification of lung cancer. In: 2023 International Conference for Advancement in Technology (ICONAT), Goa, India, pp. 1–6 (2023). https://doi.org/10.1109/ICONAT57137.2023.10080801
7. Kok, C.L., Heng, J.B., Koh, Y.Y., Teo, T.H.: Energy-, cost-, and resource-efficient IoT hazard detection system with adaptive monitoring. Sensors **25**, 1761 (2025). https://doi.org/10.3390/s25061761
8. Ali, I., et al.: Efficient lung nodule classification using transferable texture convolutional neural network. IEEE Access **8**, 175859–175870 (2020)
9. Ali, I., Muzammil, M., Haq, I.U., Amir, M., Abdullah, S.: Efficient lung nodule classification using transferable texture convolutional neural network. IEEE Access **8**, 175859–175870 (2020). https://doi.org/10.1109/ACCESS.2020.3026080
10. Dey, R., Lu, Z., Hong, Y.: Diagnostic classification of lung nodules using 3D neural networks. In: 2018 IEEE 15th International Symposium on Biomedical Imaging (ISBI 2018), Washington, DC, USA, pp. 774–778 (2018). https://doi.org/10.1109/ISBI.2018.8363687
11. Guo, N., Yen, R.-F., Fakhri, G.E., Li, Q.: SVM based lung cancer diagnosis using multiple image features in PET/CT. In: 2015 IEEE Nuclear Science Symposium and Medical Imaging Conference (NSS/MIC), San Diego, CA, USA, pp. 1–4 (2015). https://doi.org/10.1109/NSSMIC.2015.7582234
12. Kok, C.L., Ho, C.K., Chen, L., Koh, Y.Y., Tian, B.: A novel predictive modeling for student attrition utilizing machine learning and sustainable big data analytics. Appl. Sci. **14**, 9633 (2024). https://doi.org/10.3390/app14219633
13. Kaur, J., Singh, M., Singla, S.K.: Study of Fusion of medical images and classification comparison using different kernels of SVM and K-NN classifiers. In: 2020 First IEEE International Conference on Measurement, Instrumentation, Control and Automation (ICMICA), Kurukshetra, India, pp. 1–6 (2020). https://doi.org/10.1109/ICMICA48462.2020.9242697

14. Boussion, N., et al.: Contrast enhancement in emission tomography by way of synergistic PET/CT image combination. Comput. Methods Programs Biomed. **90**(3), 191–201 (2008)
15. Jakimovski, G., Davcev, D.: Using double convolution neural network for lung cancer stage detection. Appl. Sci. **9**(3), 1–12 (2019)
16. Riquelme, D., Akhloufi, M.A.: Deep learning for lung cancer nodules detection and classification in CT scans. AI **1**(1), 28–67 (2020)
17. Kok, C.L., Ho, C.K., Aung, T.H., Koh, Y.Y., Teo, T.H.: Transfer learning and deep neural networks for robust intersubject hand movement detection from EEG signals. Appl. Sci. **14**, 8091 (2024). https://doi.org/10.3390/app14178091
18. Riquelme, D., Akhloufi, M.: Deep learning for lung cancer nodules detection and classification in CT scan. In: AI, vol. 1, no. 1, pp. 28–67. MDPI AG (2020). https://doi.org/10.3390/ai1010003
19. Soundari, D.V., Poongodi, C., Sanjay Kumar, V.G., Srijith Raj, S., Srikanth, V.: Lung cancer detection using 3D-convolution neural network. In: 2022 Smart Technologies, Communication and Robotics (STCR), Sathyamangalam, India, pp. 1–5 (2022). https://doi.org/10.1109/STCR55312.2022.10009146
20. Arukonda, S., Sountharrajan, S.: Investigation of lung cancer detection using 3D convolutional deep neural network. In: 2020 2nd International Conference on Advances in Computing, Communication Control and Networking (ICACCCN), Greater Noida, India, pp. 763–768 (2020). https://doi.org/10.1109/ICACCCN51052.2020.9362857

LPDDN: An Embedded Real-Time Pavement Defect Detection Model Under Computationally Constrained Conditions

Jiaxi Guan[1], Chenxiang Li[2,3](✉), Yi Xiao[1], Yuan Xia[1], Huafang Ou[1], and Linyao Zhou[1]

[1] Hubei Institute of Measurement and Testing Technology, Wuhan, China
[2] School of Machinery and Automation, Wuhan University of Science and Technology, Wuhan, China
lichenxiang0507@163.com
[3] School of Automobile and Traffic Engineering, Wuhan University of Science and Technology, Wuhan, China

Abstract. Road transport is critical to economic development, yet pavement defects like cracks and potholes pose significant safety risks. Traditional detection methods are inefficient and resource-intensive, while existing deep learning models often fail to meet real-time requirements on edge devices. To address these challenges, we propose LPDDN (Lightweight Pavement Defect Detection Network), a novel embedded real-time detection model. LPDDN integrates a Lightweight Information Sharing Detection Head (LISD Head) to reduce model complexity, a WIoU loss function to enhance detection accuracy, and a Layer-Adaptive Magnitude-based Pruning (LAMP) method to optimize computational efficiency. Experiments on the China-MotorBike dataset demonstrate LPDDN's superior performance, achieving 94.9% mean average precision (mAP) with only 858,941 parameters. Deployment on NVIDIA Jetson Xavier NX further proves its real-time capability with detection times under 50 ms. LPDDN offers an efficient solution for real-time pavement defect detection, balancing accuracy and computational efficiency for practical applications.

Keywords: Pavement defect detection · Real-time detection · Embedded systems · Lightweight neural network

1 Introduction

As the most widely used mode of transportation, road transport plays a crucial role in the development of the national economy [1]. However, with the increasing service life of roads, both asphalt and concrete pavements inevitably develop various early-stage damages [2]. Among these, pavement cracks are particularly prominent. They are not only the earliest and most common type of pavement defect but also tend to worsen over time, posing a significant threat to traffic safety [3]. Consequently, transportation authorities worldwide consider pavement maintenance a critical task for ensuring road safety, and the timely identification of pavement cracks has become a key factor in preventing road deterioration and enhancing traffic safety [4].

L. Meng (Ed.): CCCN 2025, SCT, pp. 11–24, 2025.
https://doi.org/10.1007/978-3-032-03632-2_2

Currently, road defect detection primarily relies on two approaches: manual inspection and automated detection. Traditional manual inspection methods suffer from several limitations [5]. First, they require substantial human and material resources, resulting in low detection efficiency. Second, the results are susceptible to the expertise of inspectors, leading to subjective biases. Additionally, manual inspection processes may interfere with normal traffic flow [6]. In contrast, automated detection technologies demonstrate significant advantages [7]. Modern multifunctional road inspection vehicles are equipped with advanced sensors such as GPS positioning systems, high-definition cameras, laser profilers, and ground-penetrating radar, enabling efficient and accurate road defect detection. Since the 21st century, many countries have introduced specialized road inspection vehicles, which can perform detection tasks without disrupting traffic. Against this backdrop, the development of economical, efficient, and accurate pavement crack detection technologies holds substantial practical significance.

Traditional object detection algorithms primarily relied on handcrafted features and machine learning methods, typically involving three steps: region proposal, feature extraction, and classification with regression. For instance, Viola et al. [8] achieved efficient face detection by utilizing Haar features and AdaBoost classifiers, while the combination of Histogram of Oriented Gradients (HOG) [9] and Support Vector Machines (SVM) [10] demonstrated excellent performance in pedestrian detection. The Deformable Parts Model (DPM) [11] further improved detection accuracy by modeling the local features and spatial relationships of objects. However, these methods, due to their reliance on handcrafted features, struggled to handle complex scenarios, exhibiting limited detection accuracy and generalization capabilities, as well as low computational efficiency. With the advent of deep learning, significant progress was made in object detection algorithms. Two-stage object detection algorithms, such as the R-CNN [12] series, significantly improved detection accuracy by first generating region proposals and then classifying and regressing these proposals. Fast R-CNN and Faster R-CNN [13] further optimized speed and accuracy by introducing RoI [14] Pooling and RPNs. Mask R-CNN [15] extended Faster R-CNN by incorporating an instance segmentation branch, enabling more refined object detection. Nevertheless, the step-by-step nature of two-stage methods still resulted in relatively low computational efficiency. In contrast, single-stage object detection algorithms, such as YOLO [16], SSD [17], and RetinaNet [18], directly predicted object categories and locations in an end-to-end manner, achieving high detection speeds while maintaining competitive accuracy, making them more suitable for real-time applications. The YOLO series transformed the detection problem into a regression task, achieving extremely high inference speeds. SSD leveraged multi-scale feature maps to detect objects of various sizes, balancing detection capabilities across different scales. RetinaNet addressed the issue of class imbalance in single-stage detectors by introducing the Focal Loss, further enhancing detection accuracy. Shao et al. [19] adopted the YOLOv3 framework to propose an image processing pipeline for crack size measurement based on PTZ cameras. Mandal et al. [20] utilized YOLOv4 for crack detection, investigating the impact of various backbones. Liu et al. [21] integrated YOLOv4 and Fast R-CNN through ensemble learning. Guo et al. [22] proposed replacing the backbone network of the original YOLOv5s model with the lightweight MobileNetv3 to reduce model parameters and computational costs, while employing

Coordinate Attention to enhance model accuracy. However, these methods still struggle to fully meet the stringent requirements of low power consumption and real-time performance on edge devices, despite maintaining high accuracy.

Therefore, to address the aforementioned issues, this paper proposes a lightweight pavement defect detection model, LPDDN (Lightweight Pavement Defect Detection Network), which reduces the number of model parameters while maintaining the original accuracy, enabling it to operate more efficiently on resource-constrained embedded development boards. The main contributions of this paper are as follows:

a) The LISD Head (Lightweight Information Sharing Detection Head) is designed. By leveraging shared convolutions, the number of convolutions in the detection head is reduced. Through information sharing, the model's feature extraction capability is maintained while reducing the number of model parameters.
b) The WIoU loss function, which employs a dynamic non-monotonic focusing mechanism, is introduced as the model's loss function. By balancing the proportion of low-quality and high-quality samples, it addresses the issues of ambiguous targets and obscured pavement defects that are difficult to detect, thereby enhancing the overall performance of the model.
c) The LAMP (Layer-Adaptive Magnitude-based Pruning) pruning method is introduced for model lightweighting. An acceleration factor is designed to control the compression ratio of the model, enabling compression to any desired size.
d) The LPDDN model was successfully deployed on the NVIDIA Jetson Xavier NX development board, yielding satisfactory performance.

2 LPDDN

This study designs a novel pavement defect detection model, LPDDN, as illustrated in Fig. 1. The backbone serves as the foundational component of the model, dedicated to extracting multi-scale features from the input images. The CBS module, functioning as the initial convolutional layer, processes the input images to extract basic low-level features, establishing a foundation for higher-level feature extraction. The C3k2 module integrates multiple convolutional layers and bypass connections, facilitating efficient local feature extraction while maintaining gradient flow to enhance training stability. The SPPF (Spatial Pyramid Pooling Module) employs pooling operations at different scales, augmenting the model's ability to capture global context through multi-scale feature representation. The Cross-Stage Partial Attention (C2PSA) mechanism, a novel addition to YOLOv11, effectively filters irrelevant information, emphasizes critical features, and enhances detection capabilities in complex scenarios. The neck, acting as an intermediate layer connecting the backbone and the head, enhances object detection by fusing multi-scale features, particularly benefiting the detection of targets of varying sizes. The upsampling module increases the resolution of low-resolution feature maps, akin to sharpening blurred portions of an image, thereby improving the detection of smaller targets. The Concat module merges feature maps of different resolutions, analogous to overlaying transparent layers with varying details, enabling the model to leverage multi-level information for improved detection performance.

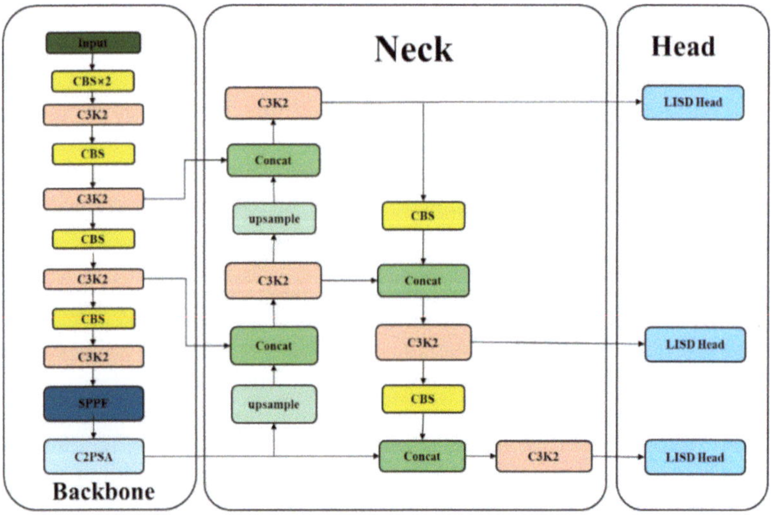

Fig. 1. LPDDN Network Structure Diagram

2.1 LISD Head (Lightweight Information Sharing Detection Head)

The three detection heads of YOLOv11 perform feature extraction through two 3 × 3 convolutions and one 1 × 1 convolution, resulting in a large number of parameters. Additionally, they predict features from only a single scale, neglecting the extraction of target features from other scales. To address this issue, this paper designs the LISD Head (Lightweight Information Sharing Detection Head), whose structure is illustrated in Fig. 2.

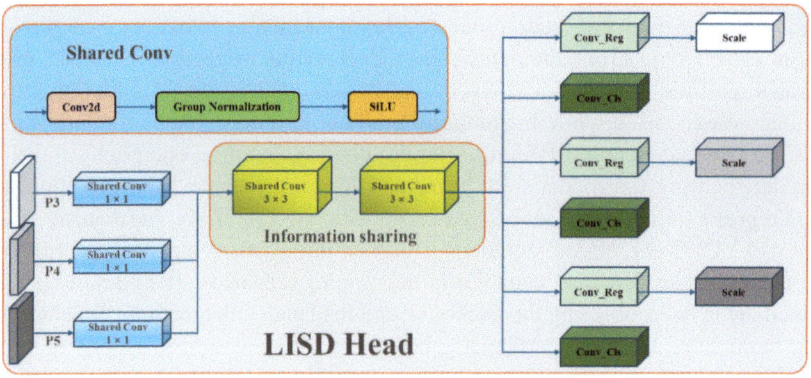

Fig. 2. Network Structure of LISD Head

This structure replaces the conventional convolution modules in the original detection heads with shared convolution modules. In the shared convolution modules, the Batch Normalization (BN) layer in the CBS convolution module is replaced with a Group

Normalization (GN) layer. The GN layer is proposed as an improvement to overcome the strong dependency of the BN layer on batch size in algorithms. The GN layer groups the feature channels and calculates the mean and variance within each group for normalization. Its computation is independent of batch size, and its accuracy remains stable across different batch sizes. Furthermore, the shared convolution strategy reduces the model's parameter count and computational complexity, making it suitable for resource-constrained devices. The LISD Head structure receives input features from the P3 to P5 levels. First, a shared convolution with a 1×1 kernel is applied to enhance information exchange across the channel dimension. Next, two shared convolutions with 3×3 kernels are used for information aggregation. Here, the two 3×3 convolutions share information, reducing redundancy and increasing the likelihood of learning adjacent feature information. Finally, the information extracted by the shared convolutions is fed into the classification and regression heads, and feature scaling is performed through the Scale layer to enhance the retention of multi-scale features. This design approach for the detection head reduces the number of convolutions in the detection heads, thereby decreasing the overall model parameter count, while maintaining the model's feature extraction capability through information sharing.

2.2 WIoU

In this study, we employ the WIoU [23] loss function, which incorporates a dynamic non-monotonic focusing mechanism, as the loss function of the model. By balancing the proportions of low-quality and high-quality samples, this loss function addresses the issues of ambiguous targets and the difficulty in detecting partially occluded objects, thereby enhancing the overall performance of the model.

The IoU-based loss, L_{IoU}, serves as the metric for measuring detection accuracy, with its calculation formula given in Eq. (1). The outlierness β is calculated according to Eq. (2) and is used to represent the abnormality of the predicted bounding box. A smaller outlierness indicates higher quality of the predicted bounding box. $\overline{L_{IoU}}$ denotes the dynamic mean loss of the predicted bounding boxes, and L^*_{IoU} is the constant derived from the transformation of $\overline{L_{IoU}}$.

A non-monotonic focusing factor r is constructed based on β as shown in Eq. (3), which reduces the model's sensitivity to loss values of easy samples and allows the model to focus more on complex samples. R_{WIoU} is used to amplify the normal-quality L_{IoU} values, with its calculation formula given in Eq. (4). Finally, the WIoU loss function is obtained as shown in Eq. (5).

$$L_{IoU} = 1 - \frac{W_i H_i}{wh + w_{gt} h_{gt} - W_i H_i} \tag{1}$$

$$\beta = \frac{L^*_{IoU}}{\overline{L_{IoU}}} \in [0, +\infty) \tag{2}$$

$$r = \frac{\beta}{\delta \alpha^{\beta - \delta}} \tag{3}$$

$$R_{WIoU} = exp\left(\frac{\left(x - x_{gt}\right)^2 + \left(y - y_{gt}\right)^2}{W_g^2 + H_g^2}\right) \in [1, e) \tag{4}$$

$$L_{WIoU} = rR_{WIoU}L_{IoU} \tag{5}$$

2.3 LAMP

This study introduces the Layer-Adaptive Magnitude-based Pruning (LAMP) method [24] and appropriately adjusts the pruning acceleration ratio for model lightweighting. Given the high demands for inference speed and storage in on-board road defect detection systems, especially when deployed on embedded devices, model pruning effectively reduces model size, enhances inference speed, and maintains high accuracy. LAMP pruning optimizes the model structure by adaptively evaluating and removing unimportant connections on a per-layer basis, thereby reducing redundant computations.

Traditional magnitude-based pruning typically applies the same pruning ratio to all layers. In contrast, LAMP automatically assigns different pruning ratios based on the distribution of weights in each layer. LAMP assesses the importance of weights based on their absolute values, with smaller absolute values indicating lesser contributions to the model and thus being candidates for pruning. The formula for LAMP scoring is given in Eq. (6).

$$LAMP \text{ score}(u; W) := \frac{(W[u])^2}{\sum_{v \geq u}(W[v])^2} \tag{6}$$

In the formula, $W[u]$ represents the weight of the u-th connection. Connections with higher scores play more important roles in the model, while those with lower scores can be pruned. After pruning, the model is fine-tuned to recover any potential performance loss. The pruning process based on LAMP scores is illustrated in Fig. 3.

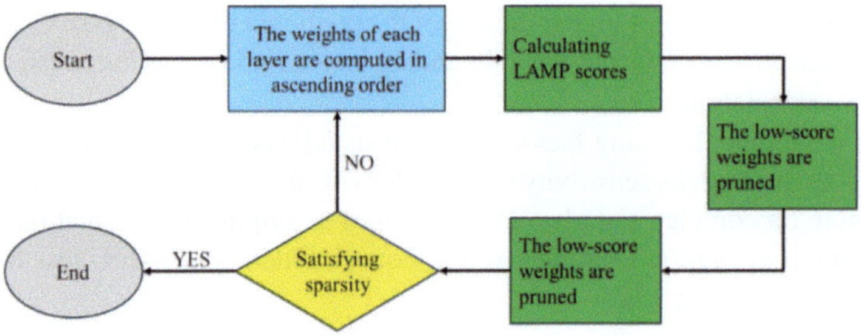

Fig. 3. The pruning process based on LAMP scores

Initially, the magnitudes of the weights in each layer are calculated and sorted in ascending order. This sorting process is crucial for accurately determining the relative importance of each weight, as it allows the model to identify and prioritize the most significant connections. Subsequently, the corresponding LAMP scores are computed based on these sorted magnitudes. Pruning is then performed by assigning zero to the elements of the mask matrix corresponding to low-scoring weights. The weight matrix is updated, and the aforementioned steps are repeated until the pre-defined sparsity level is achieved. This computational design enables the LAMP scores to more effectively measure the relative importance of channels and facilitates adaptive layer-wise sparsity. By sorting weight magnitudes in ascending order, the model can efficiently identify and remove redundant connections, thereby enhancing both computational efficiency and overall pruning effectiveness.

3 Analysis of Experimental Results

Experiments with different pruning rates were conducted to verify the model's performance under various sparsity conditions, ablation studies were conducted to verify the effectiveness of each round of improvement, and the superiority of the proposed model was demonstrated by comparing it with several existing models on the market. The experimental environment is as follows: The operating system is Windows 10, the GPU is NVIDIA GTX 2060 with a memory of 12 GB, and the version of PyTorch is 1.10.1. The total number of training epochs is 300, and the batch size for each training iteration is set to 32. The stochastic gradient descent (SGD) optimizer is employed, with the learning rate set to 0.01 in the model training hyperparameters.

This study employs the China-MotorBike subset of the ROAD2022 dataset, which focuses on motorcycle road scenarios and encompasses a variety of common road distress types. Specifically, the dataset includes five distinct labels for classifying different types of distress. These labels are as follows: D00 for longitudinal cracks, D10 for transverse cracks, D20 for alligator cracks, D40 for potholes, and Repair for areas that have been repaired. To effectively train and validate the model's performance, the dataset is divided into a training set and a testing set. The training set consists of 1,383 images, while the testing set contains 594 images.

The precision (P), recall (R), and mean average precision (mAP) are selected as the evaluation metrics for model training accuracy, while the number of parameters (Params) and floating-point operations (FLOPs) are chosen as the metrics for assessing model complexity.

Among them, precision represents the proportion of correctly predicted samples among all samples, and its calculation formula is shown in Eq. (7).

$$Precision = \frac{TP}{TP + FP} \tag{7}$$

The recall (R) represents the proportion of correctly predicted samples among all positive samples, as shown in Eq. (8).

$$Recall = \frac{TP}{TP + FN} \tag{8}$$

The calculation formula for mean Average Precision (mAP) is presented in Eq. (9). Compared to individual precision and recall, the mean Average Precision (mAP) can comprehensively evaluate the model's overall performance at different thresholds, balancing the strengths and weaknesses of precision and recall, thereby providing a more comprehensive assessment of the model's capabilities.

$$mAP = \frac{\sum_{i=1}^{N} \int_{0}^{1} P(R)dR}{N} \tag{9}$$

Here, TP denotes the number of true positives (i.e., the number of positive instances correctly classified), FP represents the number of false positives (i.e., the number of negative instances incorrectly classified), and FN indicates the number of false negatives (i.e., the number of positive instances incorrectly classified). The model parameters (Parameters) and model weight size (Size) are used to quantify the spatial complexity of the model, while the number of floating-point operations per second (Flops) is employed to measure the temporal complexity of the model.

3.1 Experiments with Different Pruning Rates

The model performance under different sp values is shown in Table 1. The sp value is used to control the compression rate of the model, and its relationship with Parameters, Flops, and size is demonstrated in Eq. (10). As sp increases from 1 to 1.5, the model's number of parameters, Flops, and size decrease progressively. However, the model's mean average precision (mAP) is highest when sp equals 1.5 and 1.75. Specifically, when sp = 1.75, the model achieves an optimal balance between accuracy and parameter quantity. Because when sp = 2, although the model's FLOPs and model parameters decrease to some extent, the accuracy also decreases. Therefore, a balance is achieved when sp = 1.75. Utilizing the LAMP method, without altering the structure of the modules, the redundant channels within the model are pruned. This process not only reduces the number of parameters from 2,420,752 to 858,941 and increases the accuracy from 94.5% to 94.9%. The reason for the improvement in mAP is that the original network had redundant channels, which actually suppressed the transmission of some information. After pruning, the number of channels and the network architecture became more compatible, thereby enhancing the mAP. Notably, LAMP achieves this by selectively removing less important channels while preserving critical information pathways, thereby enhancing the model's feature fusion capability. This selective pruning not only minimizes redundant computations but also maintains the model's inference speed by preserving essential connections.

The number of channels in each module before and after model pruning is presented in Fig. 4, where green indicates the number of channels after pruning, and orange represents the number of channels before pruning. By reducing the number of channels, the model's computational load is significantly decreased, as fewer parameters need to be processed during both training and inference. This reduction in complexity allows the model to focus on more critical features, thereby improving its learning dynamics and overall performance. Additionally, the selective pruning of channels ensures that the model retains its ability to generalize well to unseen data, maintaining robustness in diverse road defect detection scenarios.

To further clarify the influence of different pruning rates on balancing model sparsity and predictive performance in road defect detection, we observe that as the sparsity value increases, the model's sparsity improves significantly, leading to reduced computational requirements and model size. However, the mAP performance initially increases and then stabilizes, indicating that there is an optimal sparsity range where the model achieves the best trade-off between sparsity and accuracy. Beyond this range, further pruning may lead to a loss of critical information, thus negatively impacting the model's ability to detect road defects accurately. The adaptive nature of LAMP pruning allows the model to dynamically adjust its complexity based on the importance of each layer's connections, resulting in improved generalization and robustness even after extensive pruning.

$$sp \approx \frac{\text{Parameters}_{beforepruning}}{\text{Parameters}_{afterpruning}} \approx \frac{\text{FLOPs}_{beforepruning}}{\text{FLOPs}_{afterpruning}} \approx \frac{\text{Size}_{beforepruning}}{\text{Size}_{afterpruning}} \tag{10}$$

Table 1. The model performance under different sp values

sp	P	R	mAP (%)	Parameters	Flops(G)	Size(MB)
1	0.925	0.904	94.5	2,420,752	5.6	4.9
1.25	0.936	0.908	94.8	1,474,690	4.5	3.11
1.5	0.928	0.91	94.9	1,086,194	3.7	2.36
1.75	0.927	0.898	94.9	858,941	3.1	1.92
2	0.93	0.899	94.6	701,385	2.7	1.62
2.5	0.919	0.873	93.6	495,781	2.2	1.22

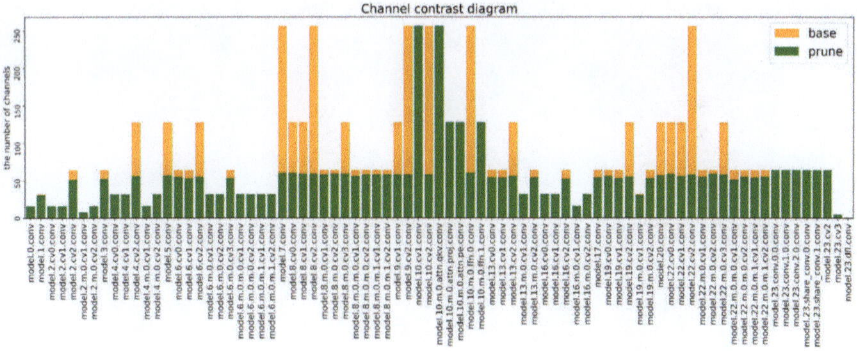

Fig. 4. The number of channels in each module before and after model pruning

3.2 Ablation Experiment

The results of the ablation studies are shown in Table 2. When the LISD Head is incorporated into the YOLO11n model, the mean average precision (mAP) of the model experiences only a marginal decrease, while the number of parameters, floating-point operations (Flops), and model size all decrease. Upon further integration of the WIOU loss, the model's mAP is improved by 0.4%. After applying LAMP pruning, the parameters, Flops, and size of the model are significantly reduced, while the mAP is enhanced to 94.9%. Compared with YOLO11n, the LPDDN model achieves a 0.6% increase in mAP, with parameters reduced to 33% of those in YOLO11n, Flops to 49%, and size to 37%. These results demonstrate that LPDDN exhibits strong feature fusion capabilities while maintaining a smaller number of parameters.

Table 2. The results of the ablation studies

model	P	R	mAP (%)	Parameters	Flops(G)	Size(MB)
YOLO11n	0.91	0.914	94.3	2,583,127	6.3	5.25
YOLO11n+L	0.917	0.907	94.1	2,420,752	5.6	4.9
YOLO11n+L+W	0.925	0.904	94.5	2,420,752	5.6	4.9
LPDDN (sp = 1.75)	0.927	0.898	94.9	858,941	3.1	1.92
LPDDN/ YOLO11n				33%	49%	37%

The detection results of pavement distress by YOLO11n and LPDDN are illustrated in Fig. 5. In Fig. 5(a), YOLO11n fails to identify the transverse crack, whereas LPDDN detects it correctly. In Fig. 5(b), for the longitudinal crack, LPDDN shows a higher confidence score than YOLO11n. In Fig. 5(c), YOLO11n fails to detect the longitudinal crack in the lower-left corner, while LPDDN identifies it correctly. In Fig. 5(d), YOLO11n fails to detect the alligator crack in the upper-left corner, whereas LPDDN detects it correctly.

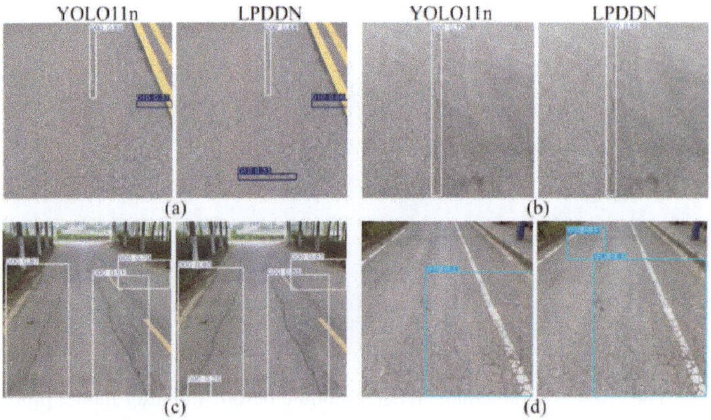

Fig. 5. The detection results of pavement distress by YOLO11n and LPDDN

3.3 Comparison Experiments of Different Models

The proposed LPDDN is compared with the existing state-of-the-art lightweight models, namely YOLOv5n, YOLOv8n, YOLOv9t, and YOLO11n. The performance comparison among different models is shown in Table 3. The mAP of YOLOv8n is 94.5%, slightly lower than that of LPDDN, which is 94.9%. However, YOLOv8n has 2,685,343 parameters, while LPDDN has only 858,941 parameters. Overall, LPDDN achieves the highest mAP while having the lowest number of parameters, flops and model size.

Table 3. The performance comparison among different models

Model	P	R	mAP (%)	Parameters	Flops(G)	Size(MB)
YOLOv5n	0.873	0.911	93.5	2,182,639	5.8	4.47
YOLOv8n	0.919	0.911	94.5	2,685,343	6.8	5.4
YOLOv9t	0.906	0.876	92.4	1,730,799	6.4	4
YOLO11n	0.91	0.914	94.3	2,583,127	6.3	5.25
LPDDN	0.927	0.898	94.9	858,941	3.1	1.92

4 Embedded Device Experiment

The LPDDN model was deployed on the NVIDIA Jetson Xavier NX, as shown in Fig. 6. The hardware parameters are detailed in Table 4, and the software information is presented in Table 5.

Fig. 6. NVIDIA Jetson Xavier NX

Table 4. Parameters of NVIDIA Jetson Xavier NX development board

Hardware name	Basic parameters
CPU	6-core NVIDIA Carmel ARM®v8.2 64-bit CPU
GPU	384-core NVIDIA VoltaTMGPU with 48 Tensor Cores

(*continued*)

Table 4. *(continued)*

Hardware name	Basic parameters
RAM	8GB 128-bit LPDDR4
ROM	16G Emmc5.1+128G SSD
USB	4xUSB 3.1,USB 2.0 Micro B
AI performance Size	21TOPS (INT8) 100 mm × 80 mm × 29 mm

Table 5. Software information

Software name	Basic parameters
Operating System	Ubuntu 20.04
Jetpack	5.1.4
OpenCV	4.5.4
CUDA	11.4
CUDNN	8.6.0
Deep Learning Environment	Pytorch

The performance of the system on the NVIDIA Jetson Xavier NX is illustrated in Fig. 7. The detection time for Fig. 7(a) is 46.56 ms, while that for Fig. 7(b) is 47.05 ms. These results demonstrate the potential for real-time pavement defect detection in vehicle-mounted systems.

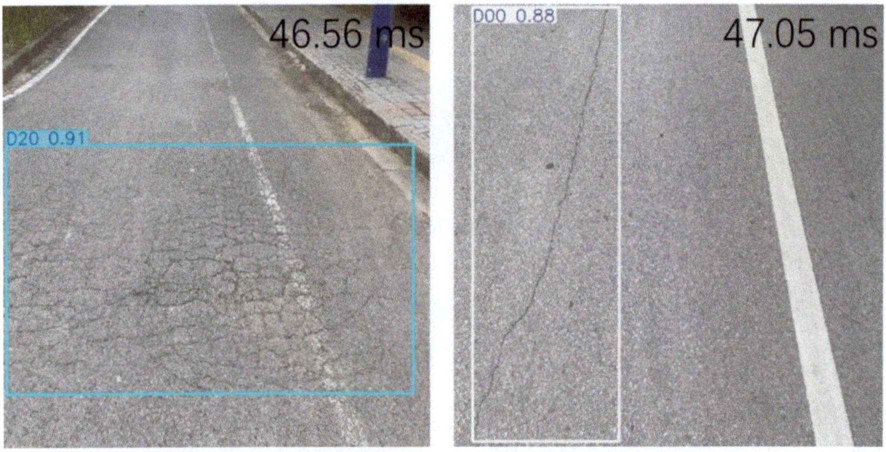

Fig. 7. The performance on the NVIDIA Jetson Xavier NX

5 Conclusion

This study presents LPDDN (Lightweight Pavement Defect Detection Network), a novel embedded real-time pavement defect detection model designed to operate efficiently under computationally constrained conditions. LPDDN addresses the critical need for accurate and timely detection of pavement defects, which is essential for maintaining road safety and preventing further road deterioration. The model's lightweight architecture and optimized components enable it to achieve high detection accuracy while significantly reducing computational resources, making it suitable for deployment on edge devices.

The primary contributions of this research include the development of the LISD Head, which reduces the number of convolutions in the detection head through shared convolutions and information sharing; the introduction of the WIoU loss function, which enhances detection performance by balancing the proportions of low-quality and high-quality samples; and the application of the LAMP pruning method, which adaptively reduces model parameters while maintaining accuracy. These innovations collectively contribute to LPDDN's superior performance compared to existing state-of-the-art models.

Extensive experiments conducted using the China-MotorBike subset of the ROAD2022 dataset demonstrate LPDDN's effectiveness. The model achieves a mean average precision (mAP) of 94.9% with only 858,941 parameters, significantly fewer than other lightweight models such as YOLOv5n, YOLOv8n, YOLOv9t, and YOLO11n. Additionally, LPDDN's deployment on the NVIDIA Jetson Xavier NX development board showcases its potential for real-time applications in vehicle-mounted systems, with detection times of 46.56 ms and 47.05 ms for different test scenarios.

In summary, LPDDN represents a significant advancement in the field of pavement defect detection, offering a practical and efficient solution for real-time monitoring and maintenance of road infrastructure. Future work may focus on further optimizing the model for broader deployment scenarios, exploring additional datasets, and enhancing the model's adaptability to diverse environmental conditions.

References

1. Fan, L., et al.: Pavement defect detection with deep learning: a comprehensive survey. IEEE Trans. Intell. Veh. **9**(3), 4292–4311 (2024). https://doi.org/10.1109/TIV.2023.3326136
2. Chen, J., Wen, Y., Nanehkaran, Y.A., Zhang, D., Zeb, A.: Multiscale attention networks for pavement defect detection. IEEE Trans. Instrum. Meas. **72**, 1–12 (2023). https://doi.org/10.1109/TIM.2023.3298391
3. Zhang, L., Xu, W., Zhu, L., Yuan, X., Zhang, C.: Study on pavement defect detection based on image processing utilizing UAV. J. Phys. Conf. Ser. **1168**(4), 042011 (2019). https://doi.org/10.1088/1742-6596/1168/4/042011
4. Luo, S., Yao, J., Hu, J., Wang, Y., Chen, S.: Using deep learning-based defect detection and 3D quantitative assessment for steel deck pavement maintenance. IEEE Trans. Intell. Transp. Syst. 1–10 (2022). https://doi.org/10.1109/TITS.2022.3169164
5. Ma, D., Fang, H., Wang, N., Zhang, C., Dong, J., Hu, H.: Automatic detection and counting system for pavement cracks based on PCGAN and YOLO-MF. IEEE Trans. Intell. Transp. Syst. **23**(11), 22166–22178 (2022). https://doi.org/10.1109/TITS.2022.3161960

6. Guo, J., Liu, P., Xiao, B., Deng, L., Wang, Q.: Surface defect detection of civil structures using images: review from data perspective. Autom. Constr. **158**, 105186 (2024). https://doi.org/10.1016/j.autcon.2023.105186
7. Liang, J., Li, G., Liu, Z.: Multi-perspective feature collaborative perception learning network for non-destructive detection of pavement defects. Digit. Sig. Process. **154**, 104661 (2024). https://doi.org/10.1016/j.dsp.2024.104661
8. Viola, P., Jones, M.: Rapid object detection using a boosted cascade of simple features. In: Proceedings of the 2001 IEEE Computer Society Conference on Computer Vision and Pattern Recognition, CVPR 2001, p. I (2001)
9. Dalal, N., Triggs, B.: Histograms of oriented gradients for human detection. In: 2005 IEEE Computer Society Conference on Computer Vision and Pattern Recognition (CVPR 2005), vol. 1, pp. 886–893 (2005)
10. Hearst, M.A., Dumais, S.T., Osuna, E., Platt, J., Scholkopf, B.: Support vector machines. IEEE Intell. Syst. Appl. **13**(4), 18–28 (1998). https://doi.org/10.1109/5254.708428
11. Felzenszwalb, P.F., Girshick, R.B., McAllester, D., Ramanan, D.: Object detection with discriminatively trained part-based models. IEEE Trans. Pattern Anal. Mach. Intell. **32**(9), 1627–1645 (2010). https://doi.org/10.1109/TPAMI.2009.167
12. Girshick, R.: Fast R-CNN. In: 2015 IEEE International Conference on Computer Vision (ICCV), pp. 1440–1448 (2015)
13. Ren, S., He, K., Girshick, R., Sun, J.: Faster R-CNN: towards real-time object detection with region proposal networks. IEEE Trans. Pattern Anal. Mach. Intell. **39**(6), 1137–1149 (2017). https://doi.org/10.1109/TPAMI.2016.2577031
14. Ding, J., Xue, N., Long, Y., Xia, G.-S., Lu, Q.: Learning RoI transformer for oriented object detection in aerial images. In: 2019 IEEE/CVF Conference on Computer Vision and Pattern Recognition (CVPR), pp. 2844–2853 (2019)
15. He, K., Gkioxari, G., Dollár, P., Girshick, R.: Mask R-CNN. In: 2017 IEEE International Conference on Computer Vision (ICCV), pp. 2980–2988 (2017)
16. Redmon, J., Divvala, S.K., Girshick, R.B., Farhadi, A.: You Only Look Once: Unified, Real-Time Object Detection. CoRR. abs/1506.02640 (2015)
17. Liu, W., et al.: SSD: single shot multibox detector. In: Leibe, B., Matas, J., Sebe, N., Welling, M. (eds.) ECCV 2016. LNIP, vol. 9905, pp. 21–37. Springer, Cham (2016). https://doi.org/10.1007/978-3-319-46448-0_2
18. Lin, T.-Y., Goyal, P., Girshick, R., He, K., Dollár, P.: Focal loss for dense object detection. In: 2017 IEEE International Conference on Computer Vision (ICCV), pp. 2999–3007 (2017)
19. Shao, C., Zhang, L., Pan, W.: PTZ camera-based image processing for automatic crack size measurement in expressways. IEEE Sens. J. **21**(20), 23352–23361 (2021). https://doi.org/10.1109/JSEN.2021.3112005
20. Mandal, V., Mussah, A.R., Adu-Gyamfi, Y.: Deep learning frameworks for pavement distress classification: a comparative analysis. In: 2020 IEEE International Conference on Big Data (Big Data), pp. 5577–5583 (2020)
21. Liu, Y., Zhang, X., Zhang, B., Chen, Z.: Deep network for road damage detection. In: 2020 IEEE International Conference on Big Data (Big Data), pp. 5572–5576 (2020)
22. Guo, G., Zhang, Z.: Road damage detection algorithm for improved YOLOv5. Sci. Rep. **12**(1), 15523 (2022). https://doi.org/10.1038/s41598-022-19674-8
23. Cho, Y.-J.: Weighted Intersection over Union (wIoU) for evaluating image segmentation. Pattern Recogn. Lett. **185**, 101–107 (2024). https://doi.org/10.1016/j.patrec.2024.07.011
24. Lee, J., Park, S., Mo, S., Ahn, S., Shin, J.: Layer-adaptive sparsity for the Magnitude-based Pruning (2021)

Improved Small Target Detection Algorithm from the Perspective of Drones

Zaijun Wang(✉), Yizhou Qian, Lin Zhou, and Xue Wang

Key Laboratory of Flight Techniques and Flight Safety, Civil Aviation Flight University of China, Deyang, China
tiantian20030315@126.com

Abstract. Compared with traditional target detection research, target detection from the perspective of drones faces more challenges, such as the constantly changing scene background, small size and unclear features of ground objects, and large and irregular movement of targets from a high altitude perspective. This makes target detection research based on drone platforms more complex and challenging than traditional target detection research. In response to the above problems, this paper proposes a target detection method for drones, which can achieve real-time, effective and accurate detection of small targets in drone scenes, and has certain performance improvements compared to existing detection methods.

Keywords: Small targets · drones · target detection

1 Introduction

Object detection is to further determine the location information of the object of interest based on image segmentation. The field of object detection has experienced the development from object detection algorithms based on manual feature construction to object detection algorithms based on deep neural networks. Before the rise of deep learning, traditional object detection algorithms were mainly based on manually designed features to generate, extract and classify target candidate regions [1]. For example, the classic scale-invariant feature transform [2] (SIFT) describes the local features of the target by calculating the gradient and direction information of key points. However, these manually designed features can only describe low-level visual information. The robustness and accuracy of extracting target features in more complex scenes are limited, and the generalization ability is insufficient, which limits the accuracy of the target detection algorithm.

In recent years, the development of deep learning neural networks has greatly promoted the progress in the field of object detection. The convolutional neural network [3] (CNN) proposed by Professor LeCun and his team has been widely used in the field of computer vision and has achieved great success in object detection tasks. GIRSHICK et al. proposed R-CNN [4], which extracts image features through CNN to detect candidate targets and inputs feature information into a fully connected network to locate the specific location of the target. IRSHICK et al. proposed Fast R-CNN [5] based on

L. Meng (Ed.): CCCN 2025, SCT, pp. 25–33, 2025.
https://doi.org/10.1007/978-3-032-03632-2_3

R-CNN, replacing the original backbone network of R-CNN with VGG-16, and then using the SS (Selective Search) method to extract candidate regions. Compared with the previous R-CNN model, Fast R-CNN has significantly improved the detection accuracy and speed, but the step of generating and screening candidate boxes is computationally intensive and the algorithm has low real-time performance. In order to achieve faster target detection, REDMON et al. proposed YOLO [6], which directly predicts the location and category of objects on the original image without screening candidate regions, greatly improving the detection speed of the model. Based on the YOLO model, REDMON et al. proposed the YOLOv2 and YOLOv3 models [7], which improved the original feature extraction network and greatly increased the detection speed of the algorithm; Bochkovskiy et al. proposed YOLOv4 [8], which innovatively improved the accuracy of YOLO series detection through a variety of data processing techniques and optimization strategies; the YOLO series has been continuously developing [9], but subsequent updates of the YOLO algorithm have mostly continued the improvement ideas of the previous generations of YOLO algorithms, mainly optimizing certain specific aspects such as training strategies and network structures, in an effort to achieve improvement and balance in the speed and accuracy of the algorithm. The newer YOLO series is the YOLOv8 model proposed by Ultralytics based on the YOLOv5 network, which has further improved detection performance and flexibility. Therefore, this detection model selects YOLOv8 as the benchmark detection model.

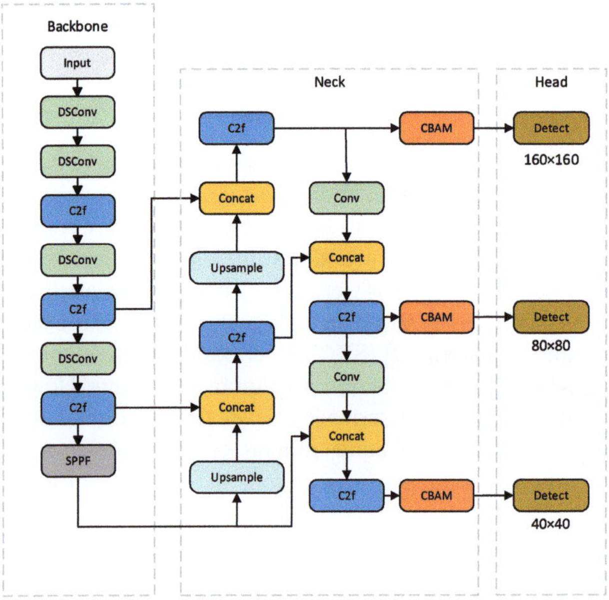

Fig. 1. Improved YOLOv8 network structure

Since the targets in aerial images are small, have unclear features, and are prone to occlusion, the detection accuracy of small targets from the perspective of drones is low, and it is easy to miss detections. Considering the poor detection effect of small targets, this paper improves the algorithm's feature extraction ability for small targets by adjusting the size of the feature map in the target detection layer. In order to address the problem of unclear target features in aerial images, this paper adds an attention mechanism after the three feature maps of the algorithm's neck network to reduce the interference of the background in aerial images on feature extraction. Finally, the ordinary convolution in the backbone network is replaced with a deep separation convolution, which reduces the number of network parameters and the amount of calculation. The overall network structure of the algorithm in this paper is shown in the Fig. 1.

2 System Description

2.1 Introducing Small Target Detection Layer

Most targets in aerial images are relatively small in size. The original YOLO series detection layer is prone to target loss or confusion when extracting small target features, and the feature extraction capability is seriously insufficient. For example, when the input image size is 640 × 640, the corresponding detection feature map sizes are 80 × 80, 40 × 40, and 20 × 20, respectively, which can be used to detect small, medium, and large targets with pixel sizes of 8 × 8, 16 × 16, 32 × 32 and above in the image. However, the size of most targets in the aerial perspective is less than 8 × 8, which means that these targets can only be assigned to one feature point for detection. At this time, the original small target detection layer cannot effectively detect smaller targets in the image, resulting in smaller targets in aerial images being easily ignored. In response to the above problems, this paper considers adjusting the size of the feature map in the small target detection layer to alleviate the problem of low accuracy of small target detection in drone aerial images. On the basis of the original target detection layer, a small target detection layer with a feature map size of 160 × 160 is added. At this time, each feature point of the detection layer corresponds to a 4 × 4 pixel area in the original image. When the size of the target is less than 8 × 8 pixels, it can still be detected by multiple feature points, that is, when the input image size is 640 × 640, the feature map size is 160 × 160, and the size of the corresponding detected target is 4 × 4 pixels. This means that multiple feature points will jointly predict the same small target, thereby improving the accuracy and robustness of the algorithm detection. At the same time, the detection layer with a feature map size of 160 × 160 only needs to undergo a small amount of convolution processing, so that a large number of underlying features can be retained, and the underlying features mainly contain fine-grained information, which is very important for positioning tasks.

At present, the conventional method is to directly add a feature map of size 160 × 160 on the basis of the feature maps of size 80 × 80, 40 × 40, and 20 × 20 in the FPN structure, that is, to directly add a small target detection layer to expand the feature map. Although it can improve the detection effect of small targets in aerial images, the existence of four detection layers and related changes will increase the number of parameters of the model. The detection layer with a feature map size of 20 × 20 is obtained by downsampling the input image 32 times, and has a large receptive field, but the pixel proportion of small targets is small. Therefore, this detection layer mainly detects targets of larger size, and the detection effect of small targets is not good.

Considering that the proportion of targets in aerial images relative to the original image is generally less than 5%, the pixel proportion of targets in the original image is not high. At this time, the cost-effectiveness of retaining the detection layer with a feature map size of 20 × 20 is very low. In addition, the detection layer with a feature map size of 20 × 20 requires a large number of convolution operations, which will cause the loss of fine-grained information of the target.

According to the above analysis, there are certain defects in directly adding a small target detection layer. This paper simplifies the model structure of the YOLOv8 network, and adjusts the size of the feature map in its small target detection layer without increasing the number of target detection layers and modules. The backbone network still uses the CSPDarknet53 structure as the feature extraction network, but removes the last convolution layer in CSPDarknet53, and also removes the C2f layer used for feature extraction and fusion, and moves the subsequent SPPF module forward. These operations result in the removal of the feature map of size 20 × 20 in the backbone network. As shown in Fig. 3.2, the sizes of the feature maps output by the streamlined backbone network are 160 × 160, 80 × 80, and 40 × 40, respectively, and the feature maps input to the neck network have also changed, so the connection position between the feature map in the FPN structure and the backbone network needs to be adjusted accordingly. The feature map of size 80 × 80 in the FPN structure is upsampled and fused with the feature map of size 160 × 160 in the backbone network, thus obtaining a detection layer with a feature map size of 160 × 160. After the streamlining operation, the feature map size of the small target detection layer also changes from 80 × 80 to 160 × 160. Since the backbone network removes the feature map of size 20 × 20, the feature map size of the large target detection layer also changes to 40 × 40. In short, by adjusting the size of the feature map in the small target detection layer, no structure is added to the model. At the same time, the redundant modules in the network are removed, the amount of training parameters is reduced, and the network structure becomes more lightweight. In addition, this paper only removes the convolutional layer of size 20 × 20 and the C2f layer behind it in the backbone network, and no longer removes the structures in the original backbone network. If too much structure of the backbone network is removed, the number of network layers will be too shallow, resulting in the inability to fully capture the feature information of different scales in the image. At the same time, the output feature map will also contain noise, making it impossible for the backbone network to fully extract the feature information of the target, which will have a certain impact on the detection effect.

2.2 Maintaining the Integrity of the Specifications

In target detection, as the number of convolutional neural network layers increases, the number of model parameters and the amount of computation required also increase. In the multi-target tracking task, the main computing resources need to be allocated to the tracking module, which leads to very limited computing resources for target detection in the multi-target tracking task. Considering the above problems, in order to better realize the multi-target tracking task from the air-to-ground perspective of drones, this paper chooses to introduce deep separable convolution [33] to achieve lightweight improvement of the YOLOv8 backbone network.

Deep separable convolution consists of two stages: depthwise convolution and pointwise convolution [10]. Its design goal is to decouple spatial filtering and channel fusion to achieve efficient computing [11]. Deep convolution maintains a high degree of independence when processing input channel data. Each input channel is assigned an independent convolution kernel. These convolution kernels slide on their respective input channels and perform convolution operations. Therefore, each input channel will generate a corresponding output channel. These forced channel independence methods maintain the spatial dimension and feature diversity of the input, and each channel is the result of independent processing. Since each channel is filtered independently, each convolution kernel only focuses on the feature extraction of a single channel, avoiding cross-channel parameter coupling.

This paper uses deep separable convolution to replace the Conv modules of layer 0, layer 1, layer 3, and layer 5 in the YOLOv8 backbone network. After replacing the standard convolution, the output result processed by the deep separable convolution module will be used as the input of the C2f module for subsequent feature extraction and fusion.

2.3 Adding Attention Mechanism

Replacing the conventional convolutional module with a lightweight convolutional module in the backbone network can effectively reduce the number of model parameters and improve computational efficiency, but this process also brings about the problem of reduced accuracy. In addition, the image background under the perspective of drone aerial photography is often complex and changeable, resulting in a large amount of spatial and channel information in the feature map generated by the model, but different spatial and channel information contribute differently to the target expression. The complex and changeable background information in the feature map will affect the target detection effect. Therefore, it is necessary to selectively allocate the feature map weights in the spatial and channel dimensions to improve the detection accuracy of the model.

The feature extraction and fusion network of the original YOLOv8 cannot fully obtain deep visual information, and the discriminability of feature representation is low, which leads to the low accuracy of the YOLOv8 model when dealing with complex background changes. CBAM is a lightweight attention mechanism module that generates feature weight coefficients in the two dimensions of channel and space through pooling operations, and multiplies them with the input feature map to readjust the weights of the feature map between the pixel level and the channel. In the YOLOv8 network, the input image is processed by the feature extraction and fusion network to obtain three feature maps with a higher degree of abstraction. At this time, the image features are fully extracted and fused. Therefore, this paper adds a CBAM module after the three feature maps obtained by the feature extraction and feature fusion network, that is, before the three detection heads, to assign weights to the feature maps in the channel and spatial dimensions, so as to obtain feature maps that pay more attention to the channel and spatial information of interest, so as to improve the detection performance of the model and improve the problems of complex background interference and occlusion.

The Convolutional Block Attention Module [12] (CBAM) is a simple and effective feedforward convolutional neural network attention module. As shown in Fig. 2, given an input feature map, CBAM will infer attention maps along the two independent dimensions of channel and space in turn, and then multiply these attention maps with the input feature map to achieve adaptive feature refinement. CBAM is a lightweight general module that can be seamlessly integrated into any convolutional neural network architecture with negligible computational overhead.

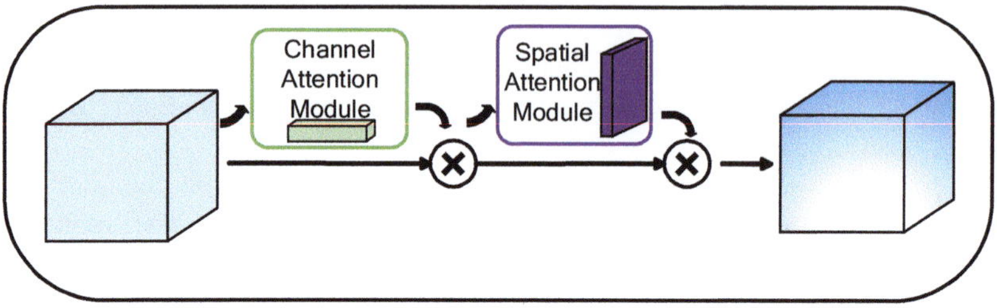

Fig. 2. CBAM Attention Mechanism

3 Experiment

3.1 Implementation Details

The deep learning framework used for model training in this article is Pytorch1.2; the computer system is Windows10; the processor model used is Intel(R) Core(TM) i5-13400F CPU @3.60 GHz; the graphics card model used is NVIDIA GeForce RTX 4060Ti; the programming language used is Python3.9; and the integrated development environment software is Pycharm2020.

3.2 Experimental Results and Analysis

In the model training stage, this paper chooses the transfer learning method to train the drone dataset. Transfer learning is a deep learning strategy that migrates the model that has been trained on a large number of datasets to our small-scale dataset for new targets, and then pre-trains the new model on the dataset, so that the final trained neural network has a good detection effect on the new target. Therefore, the pre-training weight yolov8m.pt is used as the initial network weight to train the dataset to improve the learning and optimization of the network, increase the generalization ability, and make the improved YOLOv8 model have good robustness. The training results are shown in the figure. After the training is completed, the changes in various values are basically stable (Fig. 3).

In order to verify the superiority of the improved algorithm in this paper, the YOLO series of algorithms, which are more representative in recent years, are selected for comparative experiments with the algorithm in this paper. All algorithms are trained and verified using the VisDrone2019 dataset to compare the detection effects of each algorithm on the same dataset.

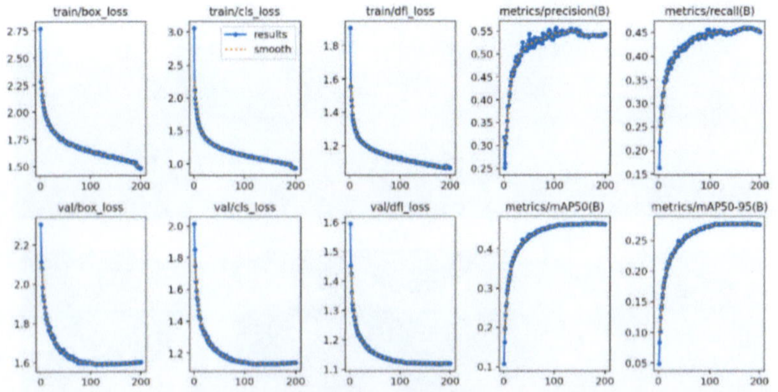

Fig. 3. Training parameter change diagram

Table 1. Comparative experimental results

Method	MAP_0.5	MAP_0.5:0.95	FPS
YOLOv5	33.7	18.2	88
YOLOv7	36.5	19.2	74
YOLOv8	39.5	23.4	118
R-CNN	33.1	16.7	15
Ours	45.0	27.6	110

From the data in Table 1, it can be seen that the YOLOv8 algorithm has certain advantages over other algorithms in the YOLO series, both in terms of detection accuracy and detection speed, so YOLOv8 is more suitable as an improved benchmark algorithm. Aiming at the problems existing in the aerial photography detection task, the corresponding method proposed in this paper is applied to the YOLOv8 algorithm. Compared with other target detection algorithms in the YOLO series, the detection accuracy is higher. From the FPS value, it can be seen that although the speed of detecting targets has decreased to a certain extent, it can also be detected in real time. Therefore, the improved YOLOv8 algorithm in this paper has better robustness for target detection in the air-to-ground perspective.

In order to intuitively verify the effectiveness of the proposed method in actual scenarios, typical images with small target size and complex road scenes in the verification set were selected, and visual comparison tests were performed using the YOLOv8 algorithm and the proposed algorithm. As shown in Fig. 4, the original YOLOv8 algorithm has missed detection and wrong detection results, while the improved algorithm is used for detection. The small targets in the figure, especially the small targets at the edge of the image, are basically all detected, ensuring that the improved algorithm model has a stable and accurate detection effect.

(a)

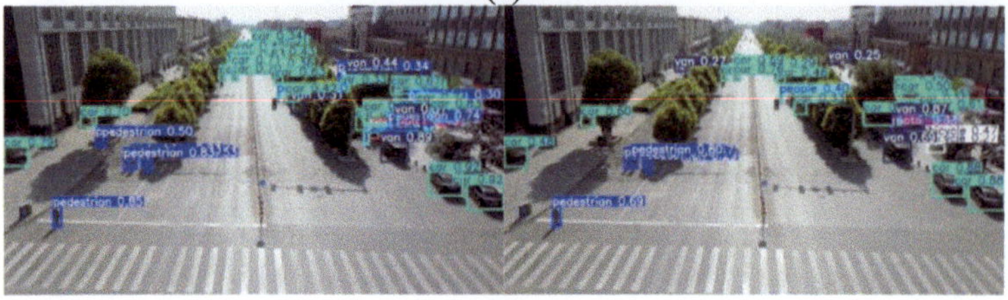

(b)

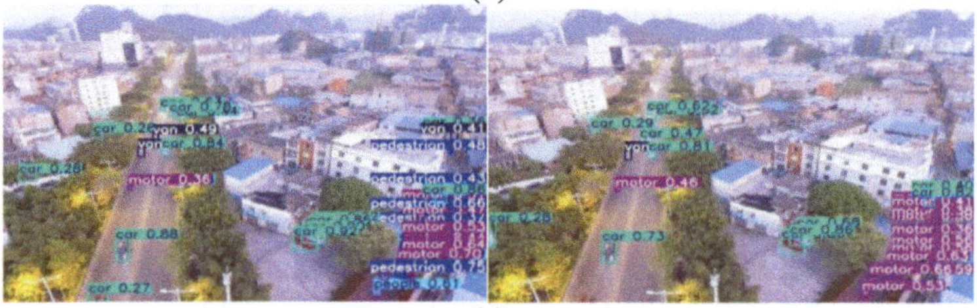

(c)

Fig. 4. Target detection result diagram. (The top is the improved detection result; the bottom is the baseline model detection result)

4 Conclusion

Aiming at the problem of poor detection of small targets from the perspective of drones, the one-stage target detection algorithm YOLOv8 model is optimized and improved. A small target detection layer is introduced on the basis of the original detection layer structure and some structural adjustments are made. The structure of the three-layer detection layer is retained to achieve more accurate detection of small targets on the ground; the CBAM attention mechanism module is added to enhance the algorithm's ability to extract feature information from feature maps of different scales; the standard convolution module in the backbone network is replaced with a depth-separable convolution module to further reduce the number of model parameters. A series of experimental results show that the improved algorithm proposed in this paper can well cope with detection tasks in complex drone scenarios.

Acknowledgement. This work was supported in part by the Independent Research Projects for Key Laboratory of Flight Techniques and Flight Safety (FZ2022ZZ02), Open Fund of Key Laboratory of Flight Techniques and Flight Safety, CAAC (FZ2022KF09) and Special project on education and teaching reform in central universities in 2023 (E2023065).

References

1. Low, D.G.: Distinctive image features from scale-invariant keypoints. J. Comput. Vis. **60**(2), 91–110 (2004)
2. Sermanet, P., Chintala, S., LeCun, Y.: Convolutional neural networks applied to house numbers digit classification. In: Proceedings of the 21st International Conference on Pattern Recognition (ICPR 2012), pp. 3288–3291. IEEE (2012)
3. Girshick, R., Donahue, J., Darrell, T., et al.: Rich feature hierarchies for accurate object detection and semantic segmentation. In: Proceedings of the IEEE Conference on Computer Vision and Pattern Recognition, pp. 580–587 (2014)
4. Ren, S., He, K., Girshick, R., et al.: Faster R-CNN: towards real-time object detection with region proposal networks. IEEE Trans. Pattern Anal. Mach. Intell. **39**(6), 1137–1149 (2016)
5. Redmon, J., Divvala, S., Girshick, R., et al.: You only look once: unified, real-time object detection. In: Proceedings of the IEEE Conference on Computer Vision and Pattern Recognition, pp. 779–788 (2016)
6. Redmon, J., Farhadi, A.: YOLOv3: an incremental improvement. arXiv preprint arXiv:1804.02767 (2018)
7. Bochkovskiy, A., Wang, C.Y., Liao, H.Y.M.: YOLOv4: optimal speed and accuracy of object detection. arXiv preprint arXiv:2004.10934 (2020)
8. Wang, A., Chen, H., Liu, L., et al.: YOLOv10: real-time end-to-end object detection. Adv. Neural. Inf. Process. Syst. **37**, 107984–108011 (2025)
9. Bewley, A., Ge, Z., Ott, L., et al.: Simple online and realtime tracking. In: 2016 IEEE International Conference on Image Processing (ICIP), pp. 3464–3468. IEEE (2016)
10. Wojke, N., Bewley, A., Paulus, D.: Simple online and realtime tracking with a deep association metric. In: 2017 IEEE International Conference on Image Processing (ICIP), pp. 3645–3649. IEEE (2017)
11. Maggiolino, G., Ahmad, A., Cao, J., et al.: Deep OC-SORT: multi-pedestrian tracking by adaptive re-identification. In: 2023 IEEE International conference on Image Processing, pp. 3025–3029 (ICIP). IEEE (2023)
12. Zhang, Y., et al.: ByteTrack: multi-object tracking by associating every detection box. In: Avidan, S., Brostow, G., Cissé, M., Farinella, G.M., Hassner, T. (eds.) ECCV 2022. LNCS, vol. 13682, pp. 1–21. Springer, Cham (2022). https://doi.org/10.1007/978-3-031-20047-2_1

Intelligent Information System and Management Based on Blockchain Integration

Blockchain-Driven Federated Learning Framework for Autonomous Medical Vehicles: Ensuring Privacy, Security, and Efficient AI Collaboration

Chiang Liang Kok(✉) and Jovan Bowen Heng

The University of Newcastle, Newcastle, Australia
{chiangliang.kok,jovanbowen.heng}@newcastle.edu.au

Abstract. Autonomous medical vehicles (AMVs) are revolutionizing emergency healthcare by enabling real-time patient monitoring, diagnostics, and decision-making. However, these systems face critical challenges such as data security, privacy, and efficient collaboration among distributed entities. This chapter proposes a novel blockchain-powered federated learning framework tailored for AMVs to address these issues. By integrating decentralized AI model training with blockchain, the framework ensures secure data sharing, robust privacy protection through advanced cryptographic techniques, and tamper-proof model updates. Smart contracts facilitate real-time model aggregation and incentivize data contribution, fostering collaboration between AMVs, hospitals, and medical networks. The proposed solution is scalable, leveraging lightweight consensus protocols, and optimized for latency-sensitive healthcare scenarios. This chapter also discusses potential applications, technical contributions, and challenges, paving the way for future advancements in autonomous medical systems.

Keywords: Blockchain · Federated learning · Autonomous Medical Vehicles (AMV) · Privacy-preserving AI · Decentralized healthcare · Smart contracts · Real-time AI collaboration · Cryptography · AI in emergency response

1 Introduction

Autonomous medical vehicles (AMVs) are transforming emergency healthcare by enabling real-time patient monitoring, diagnostics, and optimized routing. These vehicles rely heavily on artificial intelligence (AI) models trained on vast datasets collected from onboard sensors, patient records, and environmental inputs. However, integrating such systems presents significant challenges, particularly concerning data security, privacy, and decentralized collaboration among diverse medical networks [1]. Traditionally, AI systems in healthcare have made use of centralized data collection and processing architectures. While these systems have led to advancements in predictive diagnostics and decision-making, they exhibit some vulnerabilities:

- **Privacy Risks:** Centralized systems often require the transfer and storage of sensitive patient data in central repositories, increasing the risk of breaches [2].

L. Meng (Ed.): CCCN 2025, SCT, pp. 37–51, 2025.
https://doi.org/10.1007/978-3-032-03632-2_4

- **Single Points of Failure:** Centralized architectures are susceptible to network outages or cyberattacks, which can delay critical operations [3].
- **Limited Collaboration:** Medical institutions may hesitate to share data due to privacy concerns and compliance with regulations such as GDPR and HIPAA [4].

Federated learning (FL) has emerged as a promising paradigm to counter these limitations by enabling decentralized AI training. Instead of transmitting raw data to a central server, FL trains AI models locally on edge devices (e.g., AMVs) and aggregates the models to improve overall performance [5]. However, despite its privacy-preserving approach, FL introduces its own set of challenges:

1. **Trust and Transparency:** There is no guarantee that contributions from participating devices are reliable or untampered [12].
2. **Lack of Incentives:** Federated learning relies on the voluntary participation of devices, and there is no inherent mechanism to reward contributors.
3. **Data Provenance:** Ensuring the authenticity and origin of the data used for training remains a critical gap.

To address these gaps, blockchain technology has been proposed as a complementary solution to federated learning. Blockchain, with its decentralized and immutable ledger, provides many advantages:

- **Data Integrity:** Ensures tamper-proof storage of model updates and operational logs [2].
- **Incentivization Mechanisms:** Enables token-based rewards for participants contributing high-quality data or computational resources [3].
- **Privacy and Security:** Advanced cryptographic techniques, such as Zero-Knowledge Proofs (ZKPs), enhance privacy while validating data authenticity [12].

This chapter introduces a Blockchain-Driven Federated Learning Framework tailored for autonomous medical vehicles. The proposed framework leverages blockchain to address the existing limitations of federated learning, providing secure, decentralized, and privacy-preserving AI collaboration for AMVs. By incorporating lightweight consensus mechanisms and smart contracts, this framework ensures efficient real-time operations, incentivizes collaboration, and enhances trust among stakeholders in the healthcare ecosystem. In comparison to existing systems, the proposed framework represents a paradigm shift by combining the strengths of federated learning and blockchain to deliver:

- **Enhanced Security:** Blockchain ensures immutable and tamper-proof logs of AI model updates, addressing trust and transparency issues [4].
- **Scalability:** Lightweight blockchain solutions and decentralized AI training enable scalability for large networks of AMVs [12].
- **Real-Time Decision-Making:** The integration of federated learning ensures low-latency operations critical for emergency healthcare scenarios [5].

2 Literature Review

2.1 Centralized AI in Healthcare Systems

Centralized AI systems are widely implemented in healthcare for predictive diagnostics, real-time decision-making, and treatment planning. These systems aggregate vast amounts of data into centralized repositories where AI models are trained and deployed [1].

Strengths:

- Centralized systems provide high computational resources for complex model training.
- AI model management is streamlined, facilitating consistency across deployments.
- Limitations:
- Privacy Risks: Sensitive patient data is transmitted and stored in central servers, exposing it to potential breaches [2].
- Single Point of Failure: Centralized systems are vulnerable to cyberattacks, data corruption, and server outages, which can disrupt critical operations [3].
- Limited Scalability: Scaling centralized infrastructures to accommodate large datasets across institutions involves high costs and operational complexity [4].

Despite their computational benefits, centralized systems fail to address critical privacy, security, and reliability challenges, particularly in latency-sensitive healthcare applications.

2.2 Federated Learning in Healthcare and Autonomous Systems

Federated learning (FL) has come up as a decentralized AI training paradigm that enables collaboration across many institutions or devices without sharing raw data [5]. This approach is particularly important in healthcare and autonomous systems, where data privacy is paramount.

Strengths:

- **Privacy-Preserving Design:** FL eliminates the need to transmit raw data, reducing privacy risks [5].
- **Cross-Institutional Collaboration:** FL enables the use of diverse datasets, which enhances AI model robustness [6].

Limitations:

- **Trust and Transparency Issues:** FL lacks mechanisms to verify the authenticity or quality of model updates from participants, increasing the risk of data poisoning or low-quality contributions [7].
- **Incentivization Gaps:** Participants in FL networks have little motivation to contribute high-quality data or computational resources [8].
- **Data Provenance:** Existing FL frameworks do not track the origin or integrity of training data, limiting transparency [9].

While FL provides significant improvements over centralized AI, its reliance on implicit trust and lack of incentivization mechanisms hinder its adoption in critical domains like healthcare and autonomous systems.

2.3 Blockchain in Healthcare and Autonomous Systems

Blockchain technology gives promising solutions for secure data sharing, tamper-proof record management, and decentralized decision-making in healthcare and autonomous systems [6, 8, 9]. Its use in healthcare includes applications such as managing patient records, while in autonomous systems, blockchain has enabled secure vehicle-to-vehicle (V2V) communication and decentralized coordination [10].

Strengths:

- **Data Integrity:** Blockchain's immutable ledger ensures the authenticity of stored data [8].
- **Decentralized Trust:** Smart contracts enable automated operations and enforce access controls without requiring a central authority [8].

Limitations:

- **Scalability Issues:** Traditional consensus mechanisms, such as Proof of Work (PoW), are computationally intensive and unsuitable for resource-constrained devices like AMVs [11].
- **Latency Concerns:** Blockchain transaction times may not meet the real-time requirements of emergency scenarios [10].
- **Interoperability Challenges:** Blockchain integration with legacy systems and diverse platforms is complex and remains a barrier to large-scale adoption [9].

The need for scalable, low latency blockchain solutions is evident, particularly for real-time healthcare applications.

2.4 Integration of Blockchain and Federated Learning

The integration of blockchain with federated learning has been explored as a possible solution to address privacy and trust issues in decentralized AI systems [9, 11]. Blockchain's transparency and immutability enhance the reliability of federated learning networks, ensuring the integrity of model updates and incentivizing participation.

Strengths:

- **Accountability:** Blockchain tracks and verifies contributions in federated learning, ensuring transparency [8].
- **Tamper-Proof Aggregation:** Decentralized ledger technology provides tamper-proof storage of model updates, enhancing trust in the system [8].

Limitations:

- **Scalability Challenges:** Blockchain adds computational and storage overhead to federated learning networks, making it difficult to deploy at scale [10].
- **Limited Real-World Deployment:** Current studies are primarily theoretical or limited to small-scale pilots, leaving scalability and robustness questions unanswered [7].
- **Incentivization Mechanisms:** Existing frameworks rarely include dynamic reward systems to encourage high-quality contributions [11].

The proposed framework addresses these challenges by introducing lightweight consensus protocols, privacy-preserving techniques, and dynamic incentivization mechanisms tailored for AMVs.

2.5 Comparison and Rationale for the Proposed Framework

The limitations of existing approaches highlight the need for a solution that integrates the best features of blockchain and federated learning, while addressing their shortcomings. The proposed framework for AMVs offers:

1. **Privacy-Preserving AI Training:** Federated learning ensures sensitive data remains local, while blockchain validates the integrity of model updates.
2. **Enhanced Scalability:** Lightweight blockchain protocols optimize computational efficiency for resource-constrained AMVs.
3. **Dynamic Incentivization:** Smart contracts reward participants for high-quality contributions, fostering long-term collaboration.
4. **Transparency and Accountability:** Blockchain provides tamper-proof tracking of contributions and data provenance, building trust in the system.

Figure 1 below provides a comparative analysis of Centralized AI, Federated Learning, Blockchain, and the Proposed Framework, highlighting their strengths and limitations across key areas such as privacy, scalability, trust, incentivization, and real-time capability. It highlights how the proposed framework addresses critical gaps in existing systems by combining the privacy-preserving capabilities of federated learning with the trust and transparency of blockchain. This integration ensures robust privacy, scalability, and incentivization, making it particularly suitable for autonomous medical vehicles (AMVs) in emergency healthcare scenarios.

Criteria	Centralized AI	Federated Learning	Blockchain	Proposed Framework
Privacy	LOW	MODERATE	HIGH	HIGH
Scalability	MODERATE	HIGH	LOW	HIGH
Trust	LOW	MODERATE	HIGH	HIGH
Incentivization	NONE	NONE	PARTIAL	ROBUST
Real-time capability	LOW	HIGH	LOW	HIGH

Fig. 1. Comparative analysis of existing systems and proposed framework

3 Proposed Work

3.1 Introduction to the Framework

Autonomous medical vehicles (AMVs) are vital in emergency healthcare, necessitating real-time decision-making capabilities powered by advanced AI systems. However, existing approaches that rely on centralized or semi-decentralized architectures encounter difficulties such as data privacy risks, trust issues, scalability constraints, and a lack of incentivization mechanisms for collaboration. To address these challenges, this chapter proposes a blockchain-driven federated learning framework that integrates the privacy-preserving capabilities of federated learning with the trust, transparency, and incentivization potential of blockchain technology.

Key Contributions of the Framework:

1. Privacy-Preserving Model Training: Federated learning ensures that sensitive patient data remains local to AMVs, with only model updates shared on the blockchain for aggregation and validation [5].
2. Lightweight Blockchain for Scalability: The framework incorporates a resource-efficient consensus mechanism (e.g., Proof of Stake) suitable for resource-constrained environments, enhancing scalability and performance [12].
3. Dynamic Incentivization Mechanisms: Smart contracts are utilized to implement a reward-driven approach, encouraging active and honest participation from nodes in the federated learning process [13].
4. Enhanced Transparency and Trust: Blockchain provides immutable records of model contributions and data provenance, eliminating reliance on a central authority [8].

The proposed framework is particularly suited for AMVs, enabling secure and efficient collaboration across a distributed healthcare ecosystem. By addressing critical gaps in existing systems, this solution promises to enhance trust, scalability, and privacy in real-time healthcare applications. Figure 2 below shows the proposed blockchain-driven federated learning framework for autonomous medical vehicles (AMVs). The framework is made up of three core layers: the AMV Layer for data collection and local AI training using federated learning, the Blockchain Layer for validation and aggregation of federated learning model updates, and the Smart Contracts Layer for incentivization and access control. The figure also highlights the flow of federated learning model updates from AMVs to the blockchain, the reward distribution facilitated by smart contracts, and the aggregated insights provided for healthcare decision-making. This high-level overview showcases the system's ability to ensure privacy, trust, and scalability in real-time emergency healthcare scenarios.

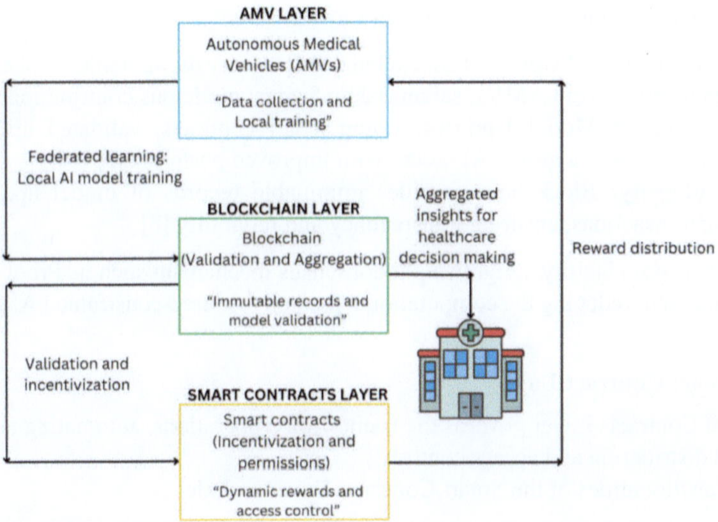

Fig. 2. Overview of the Blockchain-Driven Federated Learning Framework

3.2 System Architecture

The proposed blockchain-driven federated learning framework is composed of three primary layers, each designed to address specific challenges in privacy, scalability, and incentivization. These layers interact seamlessly to enable secure, efficient, and real-time AI-based decision-making for autonomous medical vehicles (AMVs).

3.2.1 AMV Layer

The AMV Layer operates at the network edge, comprising autonomous medical vehicles attached with sensors and computational units. This layer focuses on localized data collection and training to preserve privacy and reduce communication overhead.

Key functionalities of the AMV Layer include:

- **Data Collection:** Sensors gather real-time data such as patient vitals, environmental conditions, and vehicle diagnostics.
- **Local Model Training:** Federated learning algorithms are employed to train models locally on AMVs without transmitting raw data, thereby ensuring privacy [5].
- **Secure Model Update Transmission:** Trained model updates are encrypted and sent to the blockchain layer for validation and aggregation [12].

This layer minimizes risks to data privacy and optimizes network bandwidth by leveraging localized data processing.

3.2.2 Blockchain Layer

The Blockchain Layer serves as the framework's backbone, providing decentralized trust and security for model validation and aggregation. It ensures the integrity and authenticity of model updates through distributed ledger technology.

Main functionalities of the Blockchain Layer include:

- **Validation of Model Updates:** Blockchain nodes verify the authenticity of encrypted updates received from AMVs, safeguarding against malicious contributions [11].
- **Aggregation of Model Updates:** Using smart contracts, validated updates are aggregated to form a global AI model with improved performance [9].
- **Data Integrity:** Blockchain provides immutable records of model updates and training transactions, ensuring transparency and reliability [8].

To enhance scalability, a lightweight consensus mechanism such as Proof of Stake (PoS) is adopted, reducing the computational load on resource-constrained AMVs [12].

3.2.3 Smart Contract Layer

The Smart Contracts Layer governs the framework's operations, automating tasks such as reward distribution and access control.

Key functionalities of the Smart Contracts Layer include:

- **Dynamic Incentivization:** Smart contracts reward AMVs based on the quality of their model contributions, fostering long-term collaboration [13].
- **Access Control:** Permissions for accessing the global model or healthcare insights are managed automatically, ensuring compliance with privacy policies.
- **Scalability Support:** Incentivization mechanisms promote participation across a large number of AMVs, enabling the framework to scale effectively.

This layer streamlines operations and enhances transparency by removing reliance on centralized authorities.

Figure 3 below gives a detailed representation of the system architecture for the proposed blockchain-driven federated learning framework, highlighting the internal processes and interactions within each layer. The AMV Layer enables real-time data collection, local AI training, and emergency decision-making, with encrypted model updates sent to the Blockchain Layer. The Blockchain Layer performs critical functions, including validation, aggregation, and tamper-proof storage of model updates, supported by a lightweight consensus mechanism such as Proof of Stake for scalability. The Smart Contracts Layer automates dynamic reward distribution and access permissions management, ensuring trust and transparency without reliance on a central authority. Additionally, the system integrates a cloud-based storage solution for aggregated insights and data backup, facilitating long-term scalability and enabling real-time access for healthcare institutions. This enhanced architecture ensures privacy, security, and scalability while enabling efficient decision-making in emergency healthcare scenarios.

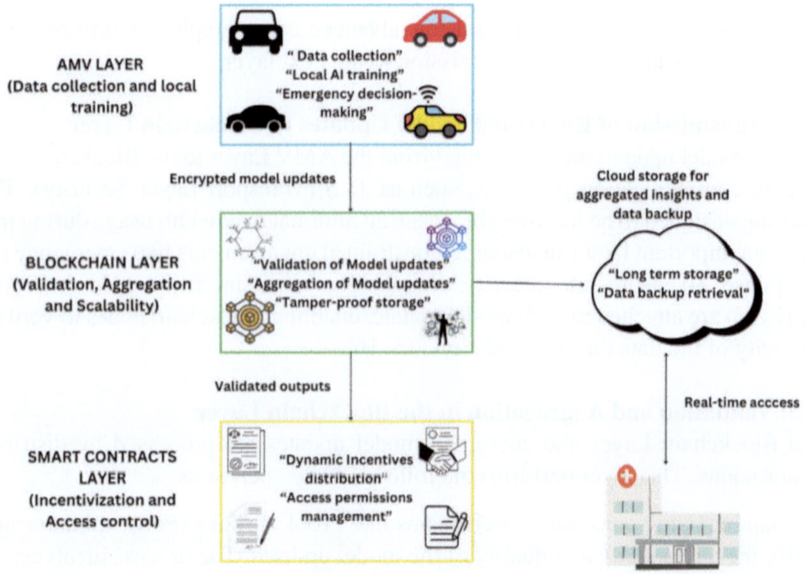

Fig. 3. Detailed system architecture of the blockchain-driven federated learning framework, illustrating key processes within the AMV, Blockchain and Smart contract layers, as well as the integration of a cloud-based storage solution for scalability

3.3 System Workflow

The proposed blockchain-driven federated learning framework provides secure, decentralized, and privacy-preserving operations for autonomous medical vehicles (AMVs). This workflow explains the step-by-step process through which data is collected, processed, validated, and utilized across the AMV Layer, Blockchain Layer, and Smart Contracts Layer, along with its integration with cloud-based systems. The technical design ensures robustness, scalability, and real-time responsiveness required for emergency healthcare environments.

Step 1: Data Collection and Local Training at the AMV Layer
At the AMV Layer, autonomous medical vehicles, equipped with advanced sensors (e.g., DHT22 for environmental monitoring, pulse oximeters for vitals), collect real-time data from their surroundings. This data includes:

- **Patient vitals:** Heart rate, oxygen levels, and body temperature.
- **Environmental metrics:** Road conditions, weather updates, and geolocation data.
- **Vehicle diagnostics:** Internal hardware health (e.g., battery status).

The data collected is processed by onboard computational units, such as NVIDIA Jetson Nano or equivalent embedded AI processors. Using federated learning algorithms, local AI models are trained directly on the AMVs. This approach eliminates the need to transmit raw patient data to a central server, ensuring compliance with privacy regulations like GDPR and HIPAA [1, 14]. The trained model updates, encapsulating learnings from the local environment, are encrypted using advanced cryptographic algorithms such as AES-256 [15], ensuring secure transmission to the next layer.

Step 2: Transmission of Encrypted Model Updates to Blockchain Layer
Encrypted model updates are transmitted from the AMV Layer to the Blockchain Layer via secure communication protocols, such as TLS (Transport Layer Security). These updates are designed to be lightweight, ensuring minimal bandwidth usage during transmission, an important factor in resource-constrained environments like emergency medical response. To ensure data integrity, digital signatures using Elliptic Curve Cryptography (ECC) are attached to each model update, enabling blockchain nodes to verify the authenticity of the data during validation [15, 16].

Step 3: Validation and Aggregation in the Blockchain Layer
At the Blockchain Layer, the encrypted model updates are processed by distributed validator nodes. This layer performs the following key operations:

1. Validation: Using consensus mechanisms like Proof of Stake (PoS), validator nodes verify the authenticity and quality of the model updates. The process involves:
   - Authenticating digital signatures using ECC.
   - Evaluating updates for consistency with previously aggregated models to detect potential anomalies or adversarial attacks [17].
2. Aggregation: Validated updates are aggregated to create a global AI model. Aggregation uses weighted averaging methods (e.g., Federated Averaging Algorithm) to combine contributions from all participating AMVs [18]. This ensures the global model reflects diverse real-world scenarios encountered by the vehicles.
3. Data Integrity: The blockchain maintains an immutable ledger of all updates, including metadata such as timestamps, contributing AMVs, and validation results. This ensures transparency and traceability of the data flow [16].

To optimize resource utilization, the framework employs sharding techniques, allowing transactions to be processed in parallel across multiple sub-chains, thereby reducing computational and storage overhead [19].

Step 4: Reward Distribution via Smart Contracts
Once the global AI model is updated, the Smart Contracts Layer automates the incentivization mechanism. Smart contracts perform the following tasks:

1. Evaluation of Contributions: Each AMV's contribution is scored based on metrics such as:
   - Model accuracy improvement (e.g., reduction in loss function).
   - Timeliness of updates (e.g., latency between data collection and transmission).
2. Dynamic Reward Allocation: Tokens or credits are distributed to AMVs proportional to their contribution scores. The framework employs Dynamic Weighted Reward Algorithms to encourage high-quality contributions while preventing adversarial behavior [17].

Smart contracts also enforce access control policies, granting permissions for healthcare institutions to access specific aggregated insights or global models. These permissions are logged on-chain to ensure compliance with privacy regulations [9].

Step 5: Aggregated Insights and Long-Term Cloud Storage
After validation and aggregation, the global AI model and aggregated insights are stored in a cloud-based system. The cloud integration serves two purposes:

1. Real-Time Access: Healthcare institutions, such as hospitals and emergency coordination centers, retrieve aggregated insights via API calls to the cloud. These insights guide real-time decisions, such as routing ambulances or adjusting emergency protocols based on current medical trends.
2. Long-Term Data Retention: The cloud serves as an off-chain storage solution for non-critical data, such as historical performance logs or aggregated statistics, ensuring scalability without overloading the blockchain network [18].

Step 6: Real-Time Decision-Making and Feedback Loops
The integrated system enables real-time decision-making by combining:

1. Local AI Models on AMVs for immediate decisions (e.g., selecting optimal routes based on patient condition).
2. Global AI Models accessed from the cloud or blockchain, ensuring that decisions are informed by aggregated insights from all AMVs.

The feedback loop ensures that global model updates are periodically pushed back to the AMVs, enabling them to refine their local models and improve their responsiveness over time. The described workflow demonstrates the technical intricacies of the blockchain-driven federated learning framework. By integrating decentralized AI training, secure validation, incentivization mechanisms, and scalable cloud storage, the framework addresses privacy, scalability, and transparency challenges in emergency healthcare. This design ensures the system remains robust, efficient, and responsive, meeting the demands of real-time medical interventions. Figure 4 below demonstrates the detailed workflow of the blockchain-driven federated learning framework for AMVs, emphasizing its iterative and secure design. Step 1 involves real-time data collection and local AI training at the AMV Layer, ensuring privacy by keeping raw data local. In Step 2, encrypted model updates, secured with AES-256 encryption and ECC signatures, are transmitted to the Blockchain Layer. Here, in Step 3, updates are validated using a Proof of Stake (PoS) mechanism and aggregated through Federated Averaging. Step 4 utilizes the Smart Contracts Layer to dynamically allocate incentives and manage access permissions. Step 5 highlights aggregated insights stored in the cloud, providing APIs for real-time healthcare access. Finally, Step 6 showcases feedback loops: validated insights from the Blockchain Layer and updated global AI models from the Cloud are fed back to the AMV Layer, ensuring real-time decision-making and continuous model improvement.

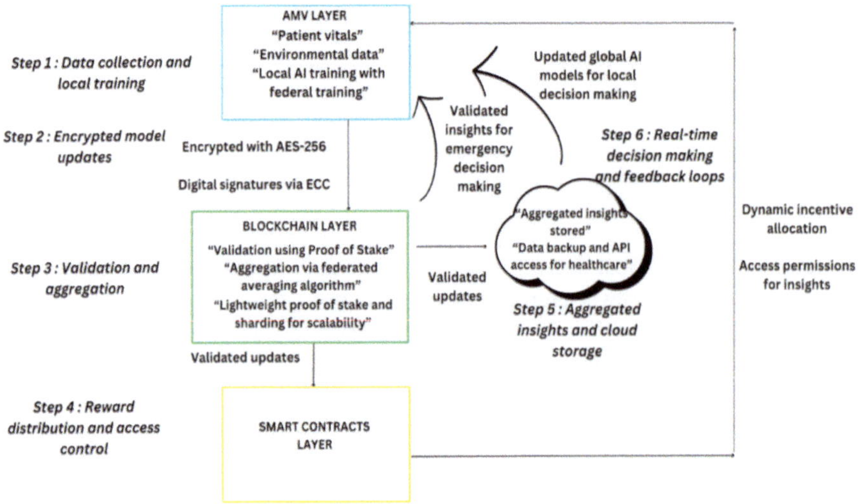

Fig. 4. Detailed workflow of the blockchain-driven federated learning framework for AMVs

Figure 5 below depicts the technical data flow and validation workflow within the proposed blockchain-driven federated learning framework. The process begins with encrypted model updates, generated at the AMV Layer, transmitted securely using AES-256 encryption and ECC digital signatures. At the Blockchain Layer, model updates undergo validation via Proof of Stake consensus and aggregation through the Federated Averaging Algorithm. Metadata such as timestamps and contributor IDs are immutably logged in the blockchain. Smart contracts automate dynamic rewards and access control, while the cloud component enables real-time API access and long-term storage. Feedback loops ensure updated global models and validated insights are periodically pushed back to the AMVs, enabling real-time decision-making and iterative model improvements.

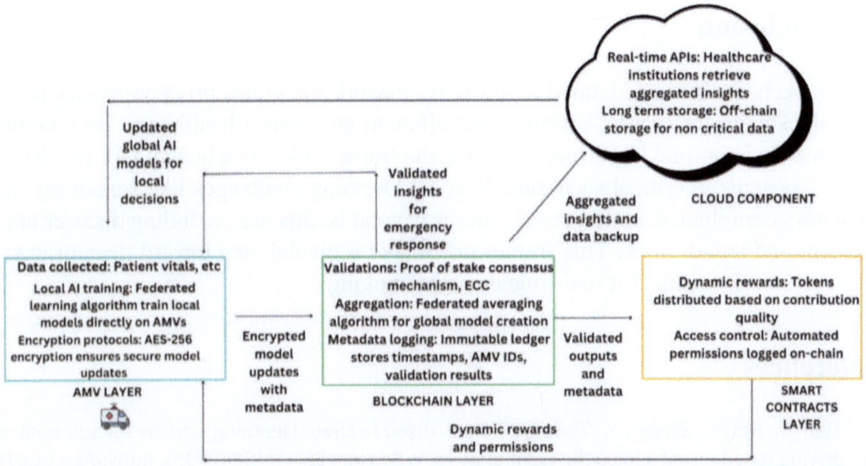

Fig. 5. Data transmission and validation workflow in blockchain-driven federated learning framework

3.4 Future Impact and Challenges

The proposed blockchain-driven federated learning framework has the potential to transform emergency healthcare systems by addressing data privacy, scalability, and trust challenges. Its integration of decentralized AI training with secure validation enables real-time decision-making for autonomous medical vehicles (AMVs). Beyond healthcare, the framework can extend to disaster management, smart cities, and IoT-based systems, providing scalable and adaptive solutions.

3.4.1 Future Impact

1. **Emergency Healthcare:** Enables faster, more informed decisions, improving patient outcomes and enhancing inter-institutional coordination.
2. **Broader Applications:** Adapts to fields like disaster management, traffic control in smart cities, and IoT systems [20] for predictive diagnostics.

3.4.2 Challenges

1. Scalability: Increased AMVs and nodes may strain blockchain capacity, requiring advanced consensus mechanisms like sharding.
2. Regulatory Compliance: Aligning with global data privacy laws like GDPR and HIPAA remains a key challenge.
3. Energy Efficiency: High energy demands of AMVs and blockchain operations necessitate optimization for sustainability.
4. Technical Limitations: Enhancing blockchain and federated learning latency is critical for real-time emergency use.

4 Conclusion

The blockchain-driven federated learning framework integrates privacy-preserving AI with blockchain for secure, scalable, and efficient emergency healthcare. By enabling real-time decision-making through AMVs, the framework sets a benchmark for decentralized systems in critical scenarios. While addressing challenges like scalability and regulatory compliance, it offers applications beyond healthcare, including disaster management and smart cities. This framework marks a pivotal step toward decentralized, trust-driven AI systems for real-time decision-making.

References

1. Tian, Y., Shi, Y., Zhang, Y., Tian, Q.: Personalized federated learning scheme for autonomous driving based on correlated differential privacy. Sensors **25**(1), 178 (2025). https://doi.org/10.3390/s25010178
2. Pokhrel, S.R., Choi, J.: Federated learning with blockchain for autonomous vehicles: analysis and design challenges. IEEE Trans. Commun. **68**(8), 4734–4746 (2020). https://doi.org/10.1109/TCOMM.2020.2990893
3. Yang, Z., Shi, Y., Zhou, Y., Wang, Z., Yang, K.: Trustworthy federated learning via blockchain. IEEE Trans. Ind. Inf. **17**(6), 3774–3785 (2021). https://doi.org/10.1109/TII.2020.3028604
4. Rieke, N., et al.: The future of digital health with federated learning. NPJ Digit. Med. **3**(1), 119 (2020). https://doi.org/10.1038/s41746-020-00323-1
5. Konečný, J., McMahan, B., Ramage, D.: Federated Optimization: Distributed Optimization Beyond the Datacenter (2015). https://arxiv.org/abs/1511.03575
6. Kok, C.L., Ho, C.K., Lee, T.K., Loo, Z.Y., Koh, Y.Y., Chai, J.P.: A novel and low-cost cloud-enabled IoT integration for sustainable remote intravenous therapy management. Electronics **13**, 1801 (2024). https://doi.org/10.3390/electronics13101801
7. Shrestha, A., Mahmood, F.: Review of deep learning algorithms and blockchain integration in autonomous systems. MDPI Sens. **22**(3), 765 (2023). https://doi.org/10.3390/s22030765
8. Smith, R.G., et al.: Distributed AI collaboration with blockchain. IEEE Access **9**, 18813–18828 (2021). https://doi.org/10.1109/ACCESS.2021.3062198
9. Zhu, T., et al.: Blockchain-federated learning: overview and challenges. J. Blockchain Res. **3**, 45–56 (2021). https://doi.org/10.1109/JBR.2021.03
10. Feng, S., et al.: Scalability solutions for blockchain in federated learning networks. J. Blockchain Technol. **4**(2), 89–102 (2022). https://doi.org/10.1109/JBT.2022.04
11. Nakamoto, S.: Bitcoin: A Peer-to-Peer Electronic Cash System (2008). https://bitcoin.org/bitcoin.pdf
12. El-Gazzar, A., Darwish, R.M., El-Mahdy, A.: Lightweight blockchain consensus mechanism and storage optimization for resource-constrained IoT devices. J. Parallel Distrib. Comput. **149**, 26–39 (2021). https://doi.org/10.1016/j.jpdc.2020.10.016
13. Feng, C., et al.: Incentive Mechanisms for Federated Learning: A Survey. arXiv preprint (2021). https://arxiv.org/pdf/2107.10243
14. Kairouz, P., et al.: Advances and open problems in federated learning. Found. Trends Mach. Learn. **14**(1), 1–210 (2021). https://doi.org/10.1561/2200000083
15. Liu, X., et al.: Secure federated learning for IoT applications using blockchain. IEEE Internet Things J. **8**(6), 4475–4488 (2021). https://doi.org/10.1109/JIOT.2020.3025433
16. Wang, H., et al.: Blockchain-based federated learning for edge computing: challenges and solutions. IEEE Netw. **35**(1), 30–36 (2021). https://doi.org/10.1109/MNET.001.2000343

17. Kang, J., et al.: Incentive mechanisms for reliable federated learning: a blockchain-based approach. IEEE Trans. Netw. Sci. Eng. **7**(3), 1397–1408 (2020). https://doi.org/10.1109/TNSE.2020.2966963
18. Zhang, Y., et al.: Federated learning meets cloud computing: opportunities and challenges. IEEE Cloud Comput. **7**(2), 21–29 (2020). https://doi.org/10.1109/MCC.2020.2972817
19. Wang, S., et al.: Sharding for blockchain scalability: opportunities and challenges. IEEE Trans. Distrib. Syst. **32**(4), 21–29 (2021). https://doi.org/10.1109/TDS.2021.3024517
20. Kok, C.L., Ho, C.K., Dai, Y., Lee, T.K., Koh, Y.Y., Chai, J.P.: A novel and self-calibrating weighing sensor with intelligent peristaltic pump control for real-time closed-loop infusion monitoring in IoT-enabled sustainable medical devices. Electronics **13**, 1724 (2024). https://doi.org/10.3390/electronics13091724

Blockchain Integration and AI-Driven Agribusiness Model Innovation for Net-Zero Transition

Fangyi Liu(✉)

National Pingtung University of Science and Technology, Pingtung 91201, Taiwan
fliu@mail.npust.edu.tw

Abstract. This research investigates the strategic integration of artificial intelligence (AI), human intelligence (HI), and blockchain technology as enablers of agribusiness model innovation (ABMI) aimed at achieving net-zero carbon transitions. Utilizing Actor-Network Theory (ANT) as the analytical framework, the study explores how these technologies function as pivotal actors to overcome traditional limitations in agribusiness, including inefficiencies in resource management, inadequate transparency, and challenges in carbon accounting. A pseudonymous Taiwanese agribusiness, referred to as "AgriNPUST," serves as the central case study to illustrate the transformative potential of these integrations. The findings outline three distinct pathways for leveraging AI and blockchain within ABMI: auxiliary acts, where AI enhances HI by automating repetitive or data-intensive tasks; autonomous acts, where AI independently manages complex operations, such as optimizing supply chain logistics; and enhanced acts, where AI, HI, and blockchain work collaboratively within decentralized ecosystems to drive innovation and sustainability. Blockchain emerges as a cornerstone, ensuring transparency, traceability, and trust through the creation of immutable records for carbon accounting and supply chain management.

A three-phase process model is proposed, comprising problem identification, interest reconstruction, and ecosystem establishment, demonstrating the synergistic role of AI and blockchain in enabling systemic innovation. The study underscores the importance of blockchain in fostering stakeholder collaboration, enhancing compliance with carbon neutrality goals, and scaling sustainable solutions. By combining theoretical insights with practical recommendations, this research highlights the role of emerging technologies in reshaping agribusiness practices.

Keywords: Blockchain Integration · Artificial Intelligence · Agricultural Innovation · Net-Zero Transition

1 Introduction

Agribusiness model innovation (ABMI) is frequently attributed to the role of managers, whose human intelligence (HI) utilizes creativity, analytical reasoning, and historical insights to respond to dynamic environmental challenges [1]. However, the inherent

L. Meng (Ed.): CCCN 2025, SCT, pp. 52–66, 2025.
https://doi.org/10.1007/978-3-032-03632-2_5

limitations of managerial expertise and access to comprehensive information, combined with the inertia of entrenched "dominant logic," often impede HI's capacity to effectively drive innovation in agribusinesses [2]. In this context, emerging technologies, particularly artificial intelligence (AI), present significant opportunities to transcend these barriers. Through the processing of extensive datasets and the disruption of conventional cognitive paradigms, AI facilitates the expansion of cognitive boundaries, thereby enhancing the potential for ABMI [3].

Despite its transformative potential, the application of AI within agribusinesses is constrained by inherent challenges, including its limited task-specific scope, lack of transparency in decision-making processes, and the substantial costs associated with its implementation and maintenance [4]. These challenges, however, can be mitigated through the strategic integration of blockchain technology, which offers enhanced transparency, traceability, and data integrity. Blockchain enables the secure recording of transactions and operations, thereby fostering trust and collaboration among stakeholders in agribusiness value chains. This synergy is particularly crucial for agribusinesses committed to net-zero transitions, as blockchain integration supports precise resource management, robust carbon accounting, and sustainable operational practices.

This study investigates the synergistic interplay of AI, HI, and blockchain technology in advancing ABMI, focusing on the case of AgriNPUST, a pseudonym for a Taiwanese agribusiness actively transitioning toward net-zero carbon goals. Leveraging ANT, which posits that both human and non-human actors participate equally in dynamic network formation and evolution, the research identifies three critical pathways for integrating AI and HI: auxiliary act (where AI assists HI in executing repetitive or routine tasks), autonomous act (where AI independently manages complex agribusiness processes), and enhanced act (where AI and HI engage in collaborative, interactive problem-solving).

AgriNPUST exemplifies these pathways by employing blockchain-integrated AI to streamline agricultural production, bolster environmental sustainability, and establish interconnected value networks that align with net-zero objectives. By seamlessly combining these technologies, AgriNPUST has successfully implemented ABMI strategies that not only advance its net-zero transition but also strengthen its competitive position in the evolving agribusiness landscape. This study offers both theoretical insights and practical guidance for agribusinesses navigating the complexities of sustainable innovation, highlighting the transformative potential of integrating AI, HI, and blockchain to achieve systemic, long-term change.

2 Theoretical Context

2.1 Agribusiness Model Innovation (ABMI)

Understanding ABMI necessitates first defining the concept of a business model (BM). A BM serves as a structured framework outlining how an agribusiness achieves its strategic objectives by creating and capturing value within an interconnected network of customers, suppliers, and partners [5]. Over the last two decades, BMs have gained prominence in strategic management, particularly as tools for addressing sustainability challenges and facilitating net-zero transitions [6].

ABMI involves the systematic transformation of BMs to address emerging challenges, such as achieving net-zero carbon emissions. This transformation requires agribusinesses to optimize resource utilization and leverage both internal capabilities and external collaborations to create and capture value throughout the agricultural value chain [7]. The essence of ABMI lies in its ability to integrate innovative activities, establish partnerships to perform specific functions, and restructure operations to align with sustainability and carbon neutrality goals [8]. Central to these efforts is the integration of blockchain technology, which enhances transparency, ensures traceability, and fosters collaboration across agribusiness value chains, enabling robust carbon accounting and emission reductions essential for achieving net-zero transitions [9].

ANT provides a comprehensive framework for analyzing ABMI by examining how "human" and "non-human" actors—such as managers, AI, and blockchain systems—interact to form, sustain, and evolve networks that address complex challenges [10]. Through the process of "translation," actors align their interests to construct dynamic and stable networks that facilitate innovative solutions. ANT thus highlights the critical interplay between technology and human decision-making in achieving net-zero objectives within agribusiness ecosystems (Leonardi, 2011).

2.2 Combining AI and HI in Enabling ABMI

According to ANT, actors encompass heterogeneous elements, both "human" and "non-human," that participate in the construction of networks by aligning their diverse interests and behavioral patterns. The principle of generalized symmetry within ANT positions human intelligence (HI) and AI as equally significant contributors to enabling ABMI [11]. AI, defined as machine intelligence designed to emulate HI, complements HI by addressing its cognitive limitations, providing advanced data processing capabilities, and enabling enhanced decision-making [12]. However, AI's inherent constraints, such as limited adaptability, algorithmic opacity, and a reliance on predefined datasets, necessitate its integration with HI to ensure effective interpretation, argumentation, and implementation [13].

The integration of blockchain technology further augments the synergy between AI and HI by offering secure, transparent, and immutable systems for data management, transaction verification, and supply chain traceability [14]. Blockchain ensures precise carbon accounting, fosters trust among stakeholders, and enhances collaboration within agribusiness value chains, making it a vital enabler of net-zero transitions [15]. By embedding blockchain capabilities, agribusinesses can track emissions reduction progress and validate sustainability practices, thereby aligning operations with carbon neutrality goals [16].

The case of AgriNPUST, a Taiwanese agribusiness engaged in a comprehensive net-zero transition, demonstrates how the combined application of AI, HI, and blockchain facilitates ABMI. AgriNPUST employs three key approaches to operationalize this synergy: (1) Auxiliary Act, where AI supports HI by automating repetitive tasks such as resource monitoring, crop health analysis, and data collection; (2) Autonomous Act, where AI independently manages complex processes, including optimizing supply chain logistics and reducing emissions; and (3) Enhanced Act, where AI and HI collaborate to develop and refine blockchain-enabled carbon accounting systems, ensuring compliance with stringent sustainability standards.

AI's evolution from a tool for specific tasks to a collaborative partner capable of advanced problem-solving underscores its transformative potential in ABMI. The integration of AI's computational capabilities, HI's contextual understanding, and blockchain's transparency offers agribusinesses like AgriNPUST a comprehensive framework for achieving innovative and sustainable transformations. This multidimensional approach not only accelerates net-zero transitions but also establishes a scalable model for systemic innovation in agribusiness, reinforcing the critical role of emerging technologies in addressing global sustainability challenges [17].

3 Methods

3.1 Research Context

AgriNPUST (a pseudonym for a Taiwanese agribusiness) serves as an exemplary case for examining ABMI through the integration of AI, HI, and blockchain technology, particularly in the context of net-zero transitions. AgriNPUST began as a traditional agricultural producer but has evolved into a dual-role organization, combining high-end agricultural systems development with sustainability solutions. This transformation highlights how technological integration can drive systemic changes in agribusiness operations and sustainability strategies.

AgriNPUST's adoption of AI and blockchain began in 2020, driven by the recognition that traditional agricultural practices were inadequate to meet modern demands for efficiency, sustainability, and carbon accountability. By introducing blockchain-integrated AI solutions for tasks such as precision crop monitoring, resource optimization, and carbon accounting, the organization has restructured its upstream and downstream agricultural value chains. In 2022, AgriNPUST launched its intelligent farm management system, combining AI's analytical capabilities with blockchain's transparency to enable real-time decision-making and traceability. Over the subsequent eight years, these innovations were scaled across all operational segments, fostering a comprehensive net-zero transition.

AgriNPUST's strategic integration of AI, HI, and blockchain exemplifies how agribusinesses can leverage emerging technologies to transform from commodity producers into sustainability-focused entities. The organization's efforts represent a significant departure from traditional practices, demonstrating the potential for ABMI to address global sustainability challenges.

3.2 Data Collection

This study employed a multi-method qualitative approach to investigate the integration of AI, HI, and blockchain in AgriNPUST's ABMI. Data were gathered through formal interviews, onsite visits, and media analysis.

Formal Interviews. Semi-structured interviews were conducted with AgriNPUST personnel, including managers from the Smart Agriculture Division and Blockchain Integration Unit, as well as engineers and researchers from the AI R&D Lab. Each interview lasted over two hours, and detailed notes were taken with participants' consent. The interviews provided insights into the technological and organizational processes underlying the integration of blockchain, AI, and HI in ABMI.

Onsite Visits. Field visits to AgriNPUST's Blockchain Research Lab, AI R&D Center, and smart agricultural production sites allowed direct observation of technological applications. These visits validated the findings from interviews and offered a clearer understanding of how blockchain-enabled systems contribute to carbon accounting and emissions reduction.

Media and Secondary Sources. Secondary data from AgriNPUST's annual reports, official announcements, and news articles (2020–2024) supplemented primary data. Keywords such as "AI," "blockchain," "net-zero transition," and "sustainable agriculture" were used to extract relevant information. These sources provided evidence of AgriNPUST's progress in achieving value creation through ABMI.

3.3 Data Analysis

The analysis followed an inductive, qualitative approach, incorporating ANT to examine the dynamic interplay between AI, HI, and blockchain in enabling ABMI.

Event History Database. Data from interviews, site visits, and secondary sources were synthesized into an event history database to document key milestones in AgriNPUST's ABMI journey.

Open Coding. Initial concepts were categorized into first-order themes, focusing on actors, blckchain-enabled transparency, and AI-HI collaboration. Coding was independently conducted by three researchers, who then cross-validated their findings to ensure consistency.

Axial Coding. First-order themes were further refined into second-order categories through iterative inductive and deductive analysis. Key themes included "blockchain-enabled traceability," "AI-driven optimization," and "collaborative ecosystem formation."

Theoretical Dimensions. Guided by ANT, the second-order themes were organized into three overarching dimensions: (1) Problem Identification, addressing inefficiencies in traditional practices; (2) Interest Reconstruction, highlighting how blockchain aligns stakeholders toward net-zero goals; and (3) Ecosystem Establishment, focusing on the integration of AI, HI, and blockchain to form sustainable networks.

Model Construction. Based on the theoretical dimensions, interim models were developed to illustrate the stages of ABMI at AgriNPUST. These models were refined iteratively and culminated in a comprehensive framework that captures the dynamic processes of technology integration in agribusiness (see Fig. 1).

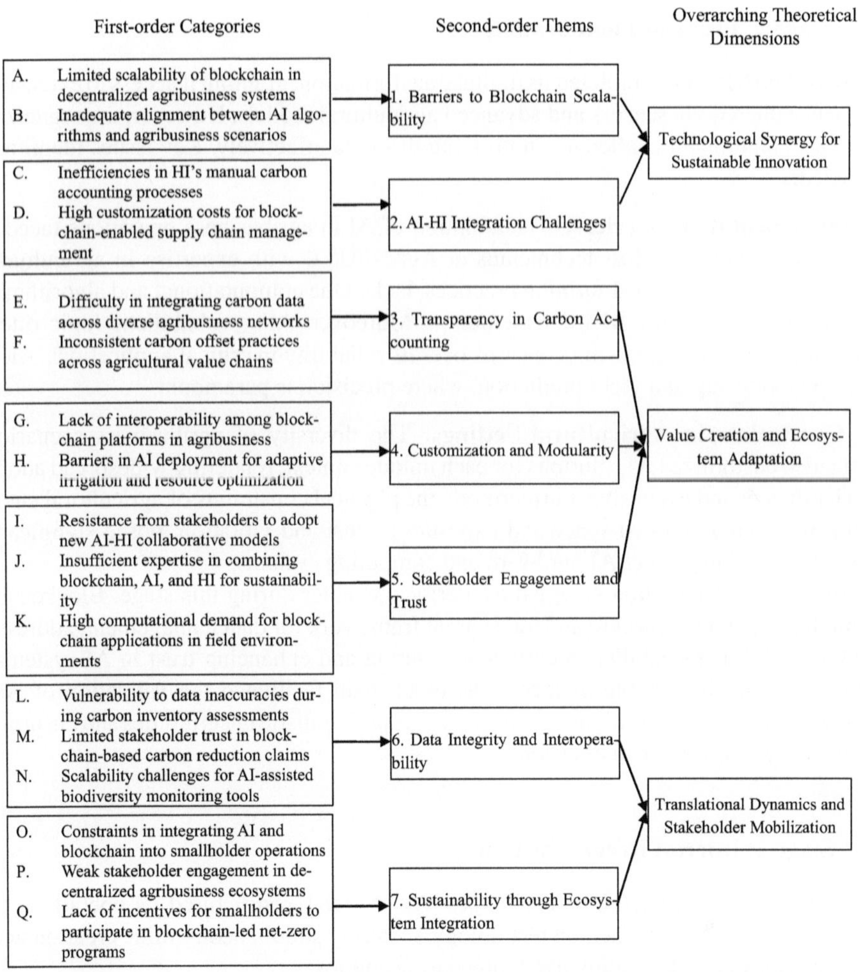

Fig. 1. Framework of Blockchain-Integrated AI and HI Synergy for Agribusiness Model Innovation

4 Findings

Our model delineates the process of agribusiness model innovation (ABMI) enabled by the integration of artificial intelligence (AI), human intelligence (HI), and blockchain technology. This process is divided into three progressive stages: problem identification, interest reconstruction, and ecosystem establishment. Below, we explore how AgriN-PUST (a pseudonym for a prominent Taiwanese agribusiness) strategically employed these technologies to advance its net-zero transition and achieve ABMI.

4.1 Stage 1: Problem Identification

In 2020, AgriNPUST completed its digital transformation, granting it access to extensive datasets gathered via sensors and advanced agricultural models. However, the organization faced significant challenges in utilizing this data effectively. Key issues identified included:

AI Implementation Barriers. The adoption of AI in agribusiness operations faced a steep learning curve. Most technicians at AgriNPUST, with expertise in agricultural engineering and traditional farming practices, lacked the computational and algorithmic knowledge required for AI implementation. Moreover, AI's probabilistic logic often introduced uncertainty, raising concerns over its reliability in critical applications, such as crop monitoring and yield prediction, where precision is paramount.

HI Constraints in Agricultural Settings. The diversity of agribusiness scenarios demanded customized AI solutions for each unique context, rendering widespread adoption inefficient and expensive. Furthermore, the physical constraints of agricultural environments, such as limited space and exposure to dust and vibration, posed significant challenges for integrating AI hardware and computing systems.

Blockchain integration emerged as a critical enabler during this stage. Blockchain technology provided a secure and transparent framework for data management, addressing the need for traceability in carbon accounting and enhancing trust in AI systems. By establishing immutable data records, blockchain facilitated the validation of AI-generated insights, thus mitigating concerns over probabilistic uncertainties and ensuring the integrity of operational decisions.

4.2 Stage 2: Interest Reconstruction

Following the identification of challenges, AgriNPUST restructured its operations to integrate AI, HI, and blockchain technology, fostering simultaneous value creation and capture. This restructuring involved three key scenarios:

Agricultural Intelligence. Blockchain-integrated AI was embedded into AgriNPUST's value chain to support HI in resource optimization and carbon footprint tracking. For instance, AI-driven models analyzed soil health and water usage, while blockchain recorded these metrics to ensure transparency and compliance with net-zero goals. This represents an auxiliary act, where AI aids HI in executing repetitive and data-intensive tasks, streamlining operational efficiency and sustainability reporting.

Intelligent Agriculturalization. AgriNPUST developed modular AI tools embedded in blockchain platforms, enabling technicians and farmers to access advanced AI functionalities without extensive technical knowledge. Customers could customize hardware solutions, such as smart irrigation systems or blockchain-enabled supply chain trackers, via an intuitive drag-and-drop interface. Blockchain ensured the traceability of agricultural inputs and outputs, fostering trust among stakeholders. This scenario illustrates an autonomous act, where AI operates independently to deliver tailored solutions, enhancing value capture through customer-centric innovations.

Network Platforms. AgriNPUST expanded its ecosystem by establishing a blockchain-based collaborative platform that connected farmers, researchers, suppliers, and other stakeholders. Blockchain technology facilitated secure data sharing across the network, enabling stakeholders to co-develop innovative solutions for sustainable agriculture. AI models optimized resource allocation and operational decisions based on real-time data from the platform. This represents an enhanced act, where AI and HI interact dynamically within a blockchain-enabled network to drive systemic innovation and achieve net-zero objectives.

Blockchain played a pivotal role in this stage by enabling the secure exchange of data across decentralized networks, ensuring transparency and fostering collaboration. The integration of blockchain further allowed AgriNPUST to create a unified framework for monitoring carbon emissions, tracking sustainability metrics, and aligning stakeholder interests around shared goals.

4.3 Stage 3: Ecosystem Establishment

The final stage involved the establishment of an integrated ecosystem, leveraging blockchain to connect internal and external stakeholders and optimize the interplay between AI and HI. Guided by Actor-Network Theory (ANT), AgriNPUST mobilized actors to form a stable and evolving network, ensuring long-term sustainability and innovation.

Internal Network. AgriNPUST created a dedicated Digital Agriculture Division to oversee the integration of AI, HI, and blockchain technologies. The division focused on developing modular blockchain-enabled AI solutions tailored to various agricultural scenarios, including precision farming, emissions monitoring, and supply chain management. A specialized AI team, trained in blockchain applications, was tasked with identifying pain points across the value chain and designing adaptive solutions.

External Network. AgriNPUST partnered with blockchain and AI technology providers to access foundational algorithms and frameworks. By incorporating blockchain technology, AgriNPUST ensured secure data exchange and transparency, enabling stakeholders to verify AI-generated insights and align operations with sustainability standards. The organization also collaborated with academic institutions and research organizations to incorporate cutting-edge blockchain and AI innovations into its ecosystem.

The blockchain-enabled ecosystem facilitated secure, transparent interactions among all participants, ensuring the traceability of agricultural practices and emissions data. This system also supported the iterative improvement of AI models through real-time feedback from network participants, enabling continuous optimization of both operational processes and carbon reduction strategies.

5 Discussion

Building on the case study of AgriNPUST's ABMI, this discussion examines how the integration of artificial intelligence (AI), human intelligence (HI), and blockchain technology facilitated the firm's transformation from a traditional agricultural producer into a dual role as a provider of sustainable agribusiness solutions and a high-end technology integrator. Using ANT, we developed a process model that highlights the dynamic interactions between AI, HI, and blockchain as key actors. This model elucidates the processes of problem identification, interest reconstruction, and ecosystem establishment, revealing their role in enabling ABMI within the context of net-zero transitions. Below, we address our research questions, theoretical contributions, practical implications, and propose avenues for future research.

5.1 Integrating AI, HI, and Blockchain in ABMI

An analysis of AgriNPUST's ABMI journey revealed how the firm employed AI, HI, and blockchain technology to achieve sustainable value creation and capture. By resolving identified challenges in combining AI and HI, AgriNPUST utilized blockchain to enhance data transparency, secure traceability, and foster trust in collaborative ecosystems. Three distinct methods of integration—**auxiliary act**, **autonomous act**, and **enhanced act**—emerged, each aligning with specific application scenarios and technical logics.

Methods of Integrating AI, HI, and Blockchain

Auxiliary Act. This foundational method involves AI assisting HI in routine agricultural tasks such as real-time crop monitoring, pest control analysis, and carbon footprint tracking. Blockchain ensures data reliability and enables seamless auditing of carbon emissions throughout the supply chain. This HI-driven approach is particularly effective in agricultural intelligence scenarios, where blockchain integrates internal and external resources to facilitate net-zero transitions through value creation.

Autonomous Act. This more advanced integration allows AI to function independently, performing complex tasks such as optimizing irrigation schedules or managing carbon trading systems without direct HI input. AgriNPUST leveraged blockchain-enabled modular AI solutions to provide customized offerings to stakeholders, enhancing transparency and reducing the operational complexities of decentralized agribusiness networks. This AI-driven approach applies to intelligent agriculturalization, where the focus is on capturing value through advanced blockchain-integrated products and services.

Enhanced Act. This most sophisticated method emphasizes iterative collaboration between AI and HI, utilizing blockchain for secure data sharing and accountability. This co-driven approach is exemplified in network platform scenarios, where AgriNPUST built an ecosystem connecting farmers, suppliers, and researchers. The platform integrates blockchain for decentralized data governance, enabling seamless collaboration and continuous optimization of solutions. This method allows for simultaneous value creation and cost reduction, advancing both technological innovation and sustainability goals.

The Process of Enabling ABMI with AI, HI, and Blockchain

Integrating ANT with ABMI, we propose a three-stage model to address the second research question: How can agribusinesses utilize AI, HI, and blockchain to enable ABMI? The model outlines the stages of problem identification, interest reconstruction, and ecosystem establishment.

Problem Identification. Aligned with ANT's problematization phase, this stage involves identifying barriers to integrating AI, HI, and blockchain in enabling ABMI. AgriNPUST identified inefficiencies in HI's capabilities and AI's probabilistic nature, particularly in managing diverse agricultural scenarios. Blockchain emerged as a critical enabler, ensuring transparent and tamper-proof data exchange, which addressed stakeholders' trust concerns and facilitated effective collaboration.

Interest Reconstruction. This stage mirrors ANT's benefitization phase, where actors redefine interests to align with the "obligatory passage points" identified in the prior stage. AgriNPUST restructured its agribusiness model around three scenarios—agricultural intelligence, intelligent agriculturalization, and network platforms—leveraging auxiliary, autonomous, and enhanced acts. Blockchain integration further reinforced these efforts by providing an immutable ledger for carbon accounting, supply chain traceability, and stakeholder engagement.

Ecosystem Establishment. Echoing ANT's recruitment and mobilization phase, this stage involves attracting internal and external actors to establish a self-sustaining ecosystem. AgriNPUST employed blockchain to enable decentralized collaboration, secure data integrity, and foster trust among participants. The resulting ecosystem supported the seamless application of AI-HI integration across all scenarios, enabling the firm to advance its net-zero objectives.

5.2 Theoretical Contributions

Integration of AI, HI, and Blockchain in ABMI

This study provides an in-depth exploration of how AI, human intelligence (HI), and blockchain technology can be synergistically integrated to drive agribusiness model innovation (ABMI). Specifically, it identifies three distinct modes of integration—auxiliary, autonomous, and enhanced acts—offering a comprehensive framework for understanding their applications in the agribusiness sector. By emphasizing the pivotal role of blockchain as a foundational enabler, this research redefines the conventional perspectives on AI-HI collaboration. Blockchain not only addresses the critical challenges of transparency and accountability inherent in ABMI but also ensures traceability across complex agribusiness value chains. For example, blockchain enables real-time verification of carbon credits and facilitates accurate recording of greenhouse gas (GHG) emissions, thus building stakeholder trust in sustainability initiatives. Through the integration of blockchain, AI's advanced analytics and HI's contextual expertise are augmented, fostering a more reliable and efficient decision-making environment, which is indispensable for achieving net-zero transitions.

A Process Model for Blockchain-Enabled ABMI

We propose a pioneering process model that merges ANT with blockchain-enabled ABMI. This model provides a structured and systematic framework for analyzing the dynamic translational actions of AI, HI, and blockchain, positioned as equal and interdependent key actors. The process elucidates how agribusinesses can leverage the complementary strengths of these actors to drive sustainable innovation. For instance, blockchain technology plays a transformative role in carbon accounting by enabling immutable, decentralized ledgers that track emissions data across various stages of the agribusiness supply chain. This ensures that the carbon footprint of agricultural practices is not only accurately measured but also transparently communicated to stakeholders. Combined with AI's ability to process vast datasets and identify optimization opportunities, such as precision farming strategies, and HI's contextual judgment in applying these insights, agribusinesses are better positioned to innovate while aligning with net-zero objectives. The model underscores the importance of blockchain in standardizing carbon management practices, enhancing accountability, and fostering collaboration across agribusiness networks, which is vital for the holistic realization of sustainability goals.

Scenario-Based Application Framework

This research further advances the theoretical understanding of ABMI by outlining a scenario-based application framework that incorporates the roles of auxiliary, autonomous, and enhanced acts across varying operational contexts. Each integration method demonstrates unique practical implications for AI-HI-blockchain collaboration within agribusiness. Auxiliary acts, for example, leverage blockchain to facilitate real-time monitoring and record-keeping for single agricultural tasks, such as tracking soil health or irrigation levels, where AI supplements HI by providing rapid, data-driven insights. Autonomous acts delve deeper into enabling blockchain-integrated AI systems to independently manage complex operations, such as optimizing resource allocation in large-scale farming networks, reducing waste, and enhancing overall efficiency without constant human intervention. Enhanced acts represent the most advanced level of integration, emphasizing continuous interaction and iteration between AI, HI, and blockchain. This approach allows for the dynamic refinement of carbon offset strategies through multi-stakeholder engagement on blockchain-powered platforms, ensuring the delivery of customized and scalable solutions that cater to diverse agribusiness scenarios.

By delineating these scenarios, this research bridges theoretical gaps and offers a practical roadmap for agribusinesses striving to balance technological innovation with sustainability. Blockchain's integration amplifies the transparency and traceability of AI-driven recommendations while reinforcing the reliability of HI-led decision-making. Such a framework not only helps agribusinesses accelerate their net-zero transitions but also positions them as leaders in sustainable agricultural practices by showcasing the potential of AI, HI, and blockchain to collaboratively achieve long-term environmental, economic, and social goals.

5.3 Practical Implications

Blockchain-Enabled Sustainability Transitions

Agribusinesses must embrace blockchain-enabled AI systems as a cornerstone for achieving net-zero sustainability transitions. By integrating blockchain with AI, these systems can provide unparalleled traceability, enabling the precise tracking of carbon emissions throughout the agribusiness value chain. Such transparency not only ensures accountability but also enhances stakeholder trust by providing verifiable data on sustainability efforts. Additionally, blockchain-based platforms streamline carbon accounting processes by recording immutable data on greenhouse gas emissions and carbon credits, ensuring compliance with increasingly stringent regulatory standards. These systems also facilitate efficient resource allocation by leveraging AI-driven insights to optimize operations, such as water usage, fertilizer application, and energy consumption, ultimately minimizing waste and maximizing productivity while adhering to sustainability goals.

Transforming Workforce Roles

The integration of AI and blockchain technologies in agribusiness necessitates a paradigm shift in workforce roles and skillsets. As these technologies permeate daily operations, agribusinesses must prioritize workforce training programs to enable seamless collaboration between human workers and intelligent systems. Employees will need to transition from performing routine tasks to focusing on strategic decision-making and oversight of AI-blockchain systems. For instance, human expertise will be essential in interpreting AI-generated insights and validating blockchain data to ensure alignment with organizational goals and sustainability standards. By fostering a technology-augmented workforce, agribusinesses can enhance operational efficiency and innovation. This transformation will not only improve productivity but also empower employees to contribute more effectively to long-term sustainability initiatives, creating a resilient and adaptive organizational culture.

Decentralized Ecosystem Development

Blockchain technology offers agribusinesses a powerful tool for developing decentralized ecosystems that drive sustainable transformation. These ecosystems integrate diverse stakeholders, including farmers, suppliers, distributors, and regulators, onto a unified platform, enabling seamless data sharing and collaborative decision-making. By leveraging blockchain's decentralized ledger, participants can access real-time, tamper-proof data on supply chain activities, resource utilization, and carbon offset initiatives, fostering collective innovation and transparency. Such ecosystems encourage the co-creation of sustainable solutions, such as shared best practices in regenerative agriculture and carbon sequestration techniques. Moreover, decentralized platforms can incentivize stakeholders to actively participate in sustainability efforts by rewarding carbon reduction milestones or sustainable farming practices with blockchain-based tokens or credits. These ecosystems not only enhance trust and cooperation among stakeholders but also position agribusinesses as leaders in driving systemic sustainability transitions in the global agricultural sector.

5.4 Future Research Directions

Deepening Synergies Between AI, HI, and Blockchain

Future research should delve deeper into the interplay between blockchain-enabled AI technologies and human intelligence (HI) to uncover additional synergies that drive innovation and sustainability in agribusiness. For instance, blockchain technologies such as smart contracts and decentralized autonomous organizations (DAOs) can automate processes and facilitate secure, transparent transactions, while HI dimensions like empathy, ethical reasoning, and contextual decision-making remain vital for addressing complex, human-centric challenges. Understanding how these advanced AI tools interact with HI in decision-making processes, particularly in areas such as sustainable resource management or ethical supply chain practices, could provide actionable insights. This exploration may reveal new frameworks for combining algorithmic precision with human intuition to achieve more effective and equitable sustainability outcomes.

Empirical Validation of the Process Model
While this study introduces a theoretical process model integrating AI, HI, and blockchain for enabling ABMI, empirical validation is essential to assess its practical applicability. Future studies should test this model across diverse agribusiness contexts, such as smallholder farming networks, large-scale agribusiness conglomerates, and cooperative organizations, to evaluate its scalability and adaptability. Comparative analyses across different regions or agribusiness types could highlight industry-specific best practices and adaptations, offering nuanced guidance for applying blockchain-enabled AI technologies to unique challenges. Additionally, longitudinal studies could provide insights into the long-term impacts of integrating these technologies on achieving net-zero transitions, improving operational efficiency, and fostering stakeholder collaboration.

Expanding Blockchain Applications in Net-Zero Transitions
Further research should investigate the broader role of blockchain in supporting agribusinesses' net-zero transitions and sustainability initiatives. Blockchain's potential extends beyond transparency and traceability to include advanced applications like carbon offset trading, biodiversity conservation, and circular economy practices. For example, blockchain could facilitate decentralized carbon credit marketplaces, allowing agribusinesses to buy and sell verified carbon offsets while ensuring integrity and avoiding double-counting. Similarly, blockchain can support biodiversity conservation by creating transparent records of regenerative agricultural practices and habitat restoration projects. In the context of the circular economy, blockchain can be utilized to track and incentivize the reuse and recycling of agricultural inputs, such as packaging materials or nutrient-rich waste. By examining these expanded applications, future studies could provide a comprehensive roadmap for leveraging blockchain as a cornerstone technology in sustainable agribusiness networks.

6 Conclusion

This study advances the understanding of ABMI by examining the synergistic integration of artificial intelligence (AI), human intelligence (HI), and blockchain technology, with a specific focus on their collective role in facilitating net-zero transitions. By employing ANT as a theoretical framework, this research conceptualizes AI and HI as equally significant actors in agribusiness networks while positioning blockchain as a pivotal enabler to overcome challenges related to transparency, accountability, and traceability. The findings underscore the transformative potential of these technologies in driving systemic innovation and operational sustainability within the agribusiness sector.

The research identifies three distinct integration pathways—auxiliary act, autonomous act, and enhanced act—each tailored to address unique challenges within agribusiness contexts. The auxiliary act focuses on leveraging AI to support HI in performing routine tasks, with blockchain enhancing data transparency and traceability. Autonomous acts extend this synergy by enabling AI to independently manage complex operations, with blockchain providing a secure, decentralized foundation for verifiable transactions and data integrity. Enhanced acts represent the highest level of integration, emphasizing iterative collaboration between AI and HI within blockchain-enabled ecosystems to co-create innovative and sustainable solutions.

The study introduces a novel process model for blockchain-enabled ABMI, encompassing three key stages: problem identification, interest reconstruction, and ecosystem establishment. Blockchain plays an integral role throughout these stages by enhancing traceability in carbon accounting, fostering stakeholder trust, and enabling collaborative decision-making within decentralized networks. This model not only provides a structured framework for understanding how agribusinesses can leverage these technologies but also serves as a roadmap for achieving sustainable innovation and advancing net-zero objectives.

The theoretical contributions of this study lie in its redefinition of AI-HI integration by incorporating blockchain as a transformative enabler. Unlike traditional perspectives that view AI as merely a supplementary tool for HI, this research positions AI, HI, and blockchain as interdependent actors capable of reshaping agribusiness practices. Furthermore, the scenario-based framework presented here bridges the gap between abstract theory and real-world applications, offering practical insights into how agribusinesses can align technological innovation with sustainability imperatives.

From a practical standpoint, this research underscores the importance of blockchain-enabled AI systems in enhancing traceability, streamlining carbon accounting processes, and fostering decentralized ecosystems for agribusiness transformation. These systems not only ensure compliance with evolving regulatory standards but also facilitate collaborative innovation across diverse stakeholder groups. Additionally, the findings highlight the necessity for agribusinesses to invest in workforce development, enabling employees to collaborate effectively with intelligent systems while focusing on strategic decision-making and oversight.

Future research directions include the empirical validation of the proposed model across diverse agribusiness contexts to refine its applicability and identify sector-specific adaptations. Moreover, further studies should explore the interplay between blockchain-enabled AI technologies—such as smart contracts and decentralized autonomous organizations—and human dimensions, including empathy, reasoning, and strategic judgment. Expanding the investigation to broader sustainability initiatives, such as carbon offset trading, biodiversity conservation, and circular economy practices, will also provide deeper insights into the role of blockchain in supporting net-zero transitions.

In conclusion, the integration of AI, HI, and blockchain technologies represents a paradigm shift in agribusiness model innovation, offering a robust framework for achieving sustainable development and global competitiveness. By addressing the unique challenges of agribusiness transformation and aligning with net-zero objectives, this study provides critical theoretical and practical insights that reinforce the role of emerging technologies in shaping the future of sustainable agribusiness practices.

References

1. Abbas, W., Zhang, Z., Asim, M., Chen, J., Ahmad, S.: AI-driven precision clothing classification: revolutionizing online fashion retailing with hybrid two-objective learning. Information **15**(4), 196 (2024)
2. Guo, F., Wu, Z., Wang, Y., Liu, C.: Analysis on the impact of dynamic innovation investment strategy of green supply chain enabled by blockchain. PeerJ Comput. Sci. **2**(April), 1–31 (2024)
3. Kumar, S., Lim, W.M., Sivarajah, U., Kaur, J.: Artificial intelligence and blockchain integration in business: trends from a bibliometric-content analysis. Inf. Syst. Front. **25**(2), 871–896 (2023)
4. Ahn, J., Yi, E., Kim, M.: Blockchain consensus mechanisms: a bibliometric analysis (2014–2024) using VOSviewer and R Bibliometrix. Information **15**(10), 644 (2024)
5. Puri, V., Mondal, S., Das, S., Vrana, V.G.: Blockchain Propels tourism industry—An attempt to explore topics and information in smart tourism management through text mining and machine learning. Informatics **10**(1), 9 (2023)
6. Edwards, M.D., Smuts, H.: Business model innovation based on disruptive technologies: a critical success factors categorization. IADIS Int. J. Comput. Sci. Inf. Syst. **19**(1), 65–80 (2024)
7. Hong, Z., Xiao, K.: Digital economy structuring for sustainable development: the role of blockchain and artificial intelligence in improving supply chain and reducing negative environmental impacts. Sci. Rep. **14**(1), 1–12 (2024)
8. Hujran, O., Alarabiat, A., Al-Adwan, A.S., Al-Debei, M.: Digitally transforming electronic governments into smart governments: SMARTGOV, an extended maturity model. Inf. Dev. **39**(4), 811–834 (2023)
9. An, S.Y., Ngayo, G., Hong, S.-P.: Enhancing 5G antenna manufacturing efficiency and reliability through blockchain and smart contract integration: a comprehensive AHP analysis. Appl. Sci. **14**(6), 2507 (2024)
10. Sun, Y., Li, L., Yu, Z., Yu, H., Wang, H.: Exploring AI models and applications within a system framework. Syst. Res. Behav. Sci. **2**(June), 1 (2024)
11. Li, S., Iqbal, M., Saxena, N.: Future industry internet of things with zero-trust security. Inf. Syst. Front. **26**(5), 1653–1666 (2024)
12. Mithas, S., Chen, Z., Saldanha, T.J.V., De Oliveira Silveira, A.: How will artificial intelligence and Industry 4.0 emerging technologies transform operations management? Prod. Oper. Manag. **31**(12), 4475–4487 (2022)
13. Periasamy, P., Dinesh, N., Padmanabhan, S.: Impact of AI in Indian BFSI sector. Optimiz.: J. Res. Manag. **16**(1), 13–17 (2024)
14. Improbable: Improbable Leverages Improved Financial Profile to Drive Innovation in AI, the Metaverse, and Web3 Through Venture Building. Business Wire (English) 1(January) (2025)
15. Innovations in Artificial Intelligence, Blockchain, Quantum Computing, Robots, and Catalysts. M2PressWIRE 2(March) (2019)
16. Lanfranchi, D., Grassi, L.: Translating technological innovation into efficiency: the case of US public P&C insurance companies. Eurasian Bus. Rev. **11**(4), 565–585 (2021)
17. Kim, Y.S., Hong, S.-P., Majer, M.: Validation of value-driven token economy: focus on blockchain content platform. Future Internet **16**(5), 178 (2024)

Cloud Computing and Intelligent Computing Model

Fourier Model Agnostic Meta-Reinforcement Learning Network

Zhe Hao[1(✉)], Zhiheng Lai[2(✉)], and Xuanyu Xue[2]

[1] Nanjing Agricultural University, Jiangpu, Nanjing 210095, Jiangsu, China
1464103497hz@gmail.com

[2] Shanghai Jiao Tong University, Jiangchuan Road, Shanghai 200030, China
{laizhihengfzyzsjtu,sjtuxxy0398}@sjtu.edu.cn

Abstract. Meta-reinforcement learning is a rapidly developing field in AI, designed to address complex application domains with similar tasks, enabling quick mastery of tasks within a specific domain through meta-learning. However, existing meta-RL algorithms generally suffer from low efficiency and are mostly applied to low-dimensional task distributions, performing poorly in high-dimensional tasks. Meanwhile, Model-Agnostic Meta-Learning (MAML) is one of the best meta methods in the meta-algorithms field. MAML is easy to deploy, widely applicable, and powerful, but it also has several issues, such as being highly sensitive to neural network architectures during training, which often leads to instability, requiring tedious hyperparameter searches to stabilize training, and incurring high computational costs during both training and inference. As the number of subtasks increases, MAML becomes difficult to generalize effectively for meta-learning, and the model tends to memorize periodic data rather than deeply understanding the underlying principles behind the periodicity. This paper proposes FMAN (Fourier Model-Agnostic Network), which aims to improve the performance and sample efficiency of meta-learning by using Fourier Transform (FT) to extract frequency-domain information from continuous space. Specifically, I extract frequency-domain information from time-series data and process it using the actor-critic algorithm to help the agent make better decisions. One of the appealing features of this network is that it can learn underlying periodic patterns without the need to predict infinitely long states. Experiments show that FMAN offers a promising solution for improving sample efficiency and generalization in meta-worlds, establishing it as a standard benchmark in meta-learning.

Keywords: Reinforcement Learning · Meta Learning · Multi-tasks Reinforcement Learning · Fourier Transform

1 Introduction

Deep reinforcement learning (DRL) has been proven to achieve significant success in complex sequential decision-making tasks, such as computer games [1], robotic

Z. Hao and Z. Lai—Equal contribution and considered co-first authors.

L. Meng (Ed.): CCCN 2025, SCT, pp. 69–81, 2025.
https://doi.org/10.1007/978-3-032-03632-2_6

control [2], and combinatorial optimization [3]. However, these algorithms often require large amounts of data for training in order to learn satisfactory control policies, which limits the applicability of DRL in real-world problems. In other words, in cases with limited data, traditional deep neural networks often fail to perform well, indicating some inherent limitations of these methods that need improvement. The most critical issue is that in many practical scenarios, data available for each class is limited, and collecting additional data is either extremely costly and time-consuming or completely infeasible. This creates a need to develop algorithms that can effectively utilize the few available samples to achieve satisfactory training results [4].

In simple terms, meta-learning is essentially a machine learning method. Its core idea is to help an agent gain more experience from a set of similar tasks, gradually improving its ability to handle new tasks. This approach allows the agent not only to solve specific tasks but also to learn how to quickly adapt when faced with new tasks, making it more similar to human learning. Therefore, meta-learning can be seen as operating on two levels: **task-level** and **meta-level**. At the **task-level**, the goal of the agent is to quickly learn the task-specific knowledge from the training data, enabling the model to adapt and solve the specific task efficiently. At the **meta-level**, the goal of meta-learning is to accumulate knowledge over time, allowing the model to learn cross-task commonalities, enabling it to retain the ability to learn and adapt when faced with unseen tasks, thus achieving effective generalization across tasks.

In recent years, meta-learning has made significant progress in multiple fields, especially in complex reinforcement learning and transfer learning problems, achieving state-of-the-art results. One particularly successful method is *Model-Agnostic Meta-Learning* (**MAML**), which has performed exceptionally well in various applications. The core idea of MAML is to find an appropriate initialization that allows the model to quickly adapt and efficiently solve new tasks with only a few training steps. In this way, MAML ensures strong generalization ability while avoiding the need for extensive training on each new task, saving time and reducing training costs.

Despite MAML achieving impressive results in many standard tests and applications [5], it still has some limitations and significant room for improvement. For example, MAML may face training instability in certain tasks, or it may struggle to fully utilize cross-task knowledge when the task differences are more significant. Therefore, further optimization of the MAML algorithm to enhance its performance in more practical applications remains an important direction for current research.

Since MAML requires updating multiple tasks during the actual training process to achieve better generalization, its performance may decline when the number of tasks is small or the task distribution is uneven. Meanwhile, the similarity between tasks also affects MAML's performance. If the tasks differ significantly, it can lead to overfitting or difficulty in quickly adapting to new tasks. Most importantly, MAML needs to simultaneously optimize the parameters of the inner and outer loops, which makes it more complex than traditional training processes. Traditional training usually only considers how to optimize the model's parameters, while MAML needs to optimize both the model's parameters and the entire training process. Therefore, to improve MAML's training efficiency, it is necessary to focus on feature processing, fully utilizing the inherent structural information in sequence state signals [6], and leveraging this information to improve long-term decision-making quality, thus overcoming the challenges brought by task differences. I believe that in reinforcement learning tasks, long-term decisions made by agents rely on accurate predictions of future states and rewards. Although existing reinforcement learning methods can already be trained using basic reward signals or supervision signals, traditional sparse reward signals and one-step prediction models often fail to capture the long-term dependencies in the environment, leading to inefficient training processes. To further improve sample efficiency and decision-making performance in reinforcement learning, I propose a method that combines the Fourier transform for state sequence prediction: SPF (State Sequences Prediction via the Fourier transform) [6]. SPF effectively extracts inherent structural information from state sequences through frequency-domain analysis, providing richer supervisory signals for long-term decision-making, which helps RL agents make more accurate decisions over longer time spans. I then designed a new network architecture–FAMN (Fourier Model Agnostic Network). The FAMN network fully utilizes the key information extracted from the frequency-domain structure and can quickly adapt in meta-learning tasks, improving sample efficiency. It uses recent policy optimization ideas for single-task adaptation, while maintaining an efficient learning process across multiple tasks. FAMN uses self-supervised tasks to predict future states through Fourier Transform, reducing the need for high-precision models while improving training stability and data efficiency.

Specifically, the FAMN I designed effectively utilizes long-term structural information in state sequences, improving reinforcement learning performance in long-term decision-making. At the same time, the FAMN network can extract patterns from periodic sequences in different tasks, which to some extent allows it to ignore the impact of task differences, enabling it to effectively use subtasks for meta-updating in multi-task meta-learning environments. This improves the performance and efficiency of the MAML algorithm to some extent. Through extensive experiments, FAMN outperforms traditional reinforcement learning algorithms in various benchmark tests in the MetaWorld tasks, demonstrating its improved learning ability in complex environments.

2 Preliminaries

2.1 Markov Decision Process

The most commonly used modeling approach in reinforcement learning problems is to frame the interaction between the agent and the environment as a Markov Decision Process [7]. The standard MDP framework [8] can be described by the tuple $M = \langle S, A, R, P, \mu, \gamma \rangle$, where the components are defined as follows: S

represents the set of all possible states, A is the set of actions that the agent can take, $R : S \times A \times S \rightarrow [-R_{\max}, R_{\max}]$ is the reward function that provides the immediate reward for transitioning between states under certain actions, $P : S \times A \times S \rightarrow [0, 1]$ denotes the transition probability function, which defines the likelihood of moving from one state to another given a specific action, $\mu : S \rightarrow [0, 1]$ is the initial state distribution that specifies the starting state of the environment, and $\gamma \in [0, 1]$ is the discount factor used to weigh future rewards.

In reinforcement learning, the agent's actions are governed by a policy π, which defines a probability distribution over possible actions given the current state, expressed as $\pi(a|s)$. The agent starts in an initial state $s_0 \sim \mu$ and at each step $t \geq 0$, the agent selects an action a_t according to the policy π (i.e., $a_t \sim \pi(\cdot|s_t)$). Once the action is taken, the environment transitions to a new state s_{t+1} based on the transition probabilities $P(s_{t+1}|s_t, a_t)$ and provides an immediate reward $r_t = R(s_t, a_t, s_{t+1})$. The goal of reinforcement learning is to identify an optimal policy π^* that maximizes the expected total future rewards.

The effectiveness of a policy π is generally assessed by the expected sum of future rewards, discounted over time. This sum, referred to as the return, represents the cumulative reward the agent anticipates when following the policy π. The return is a crucial metric for evaluating how well a policy achieves the objective of maximizing long-term rewards. The return is defined as:

$$J(\pi, M) = \mathbb{E}_{\tau \sim (\pi, M)} \left[\sum_{t=0}^{\infty} \gamma^t R(s_t, a_t, s_{t+1}) \right],$$

where $\tau = (s_0, a_0, s_1, a_1, \dots)$ denotes a trajectory generated from the interaction process and $\tau \sim (\pi, M)$ indicates that the distribution of τ depends on π and the environment model M. For simplicity, I write $J(\pi)$ and $\tau \sim \pi$ as shorthand. I also introduce the discounted future state distribution d^π, which is defined by $d^\pi(s) = (1 - \gamma) \sum_{t=0}^{\infty} \gamma^t \mathbb{P}(s_t = s|\pi)$. This allows us to express the expected discounted total reward compactly as

$$J(\pi) = \frac{1}{1-\gamma} \mathbb{E}_{s_0 \sim \mu} \mathbb{E}_{\tau \sim \pi} \left[\sum_{t=0}^{\infty} \gamma^t R(s_t, a_t, s_{t+1}) \right].$$

2.2 Meta-RL Definition

Reinforcement Learning (RL) algorithms [9] are typically created, developed, and tested by humans. Meta-RL, however, seeks to learn parts of an algorithm f through machine learning. While traditional RL focuses on learning a policy, meta-RL aims to learn the RL algorithm f that generates the policy. This approach doesn't eliminate the human role entirely but shifts it towards creating training environments and the parameters necessary for learning these algorithmic components in a data-driven manner.

Terminology: I usually parameterize f_0 with meta-parameters θ and train these parameters to maximize a meta-RL objective. As a result, f_0 directly produces the parameters of π_ϕ, i.e., $\phi = f_0(D)$. The policy π_ϕ is referred to as the base policy with base parameters ϕ. Here, D represents a meta-trajectory that might include several policy episodes. Some meta-RL methods use f_0 to map D to ϕ at each MDP step, though others do so less frequently.

2.3 MAML: Model Agnostic Meta-Learning

Many existing designs for the inner-loop algorithms, such as f_θ, are built upon established reinforcement learning (RL) techniques and incorporate elements of meta-learning to improve their performance. One prominent example in this category is the Model-Agnostic Meta-Learning (MAML) algorithm [10], which has gained significant attention. In the MAML framework, the inner-loop utilizes a policy gradient method, where the initial model parameters are treated as meta-parameters, specifically setting $\phi_0 = 0$. The key innovation in MAML lies in the inner-loop being designed as a differentiable function with respect to these initial parameters enabling the initialization to be fine-tuned through gradient-based optimization techniques, such as gradient descent [11]. This capability allows MAML to effectively serve as a robust starting point for learning tasks drawn from a broader task distribution.

When the model is tasked with adapting to a new environment, MAML first collects data using the initially provided policy [12]. Following this, the parameters are updated by performing a policy gradient step on the data collected from a task $M_i \sim \rho(M)$. This process ensures that the model is able to quickly adjust to new tasks with minimal adaptation, leveraging prior experience from the meta-learning process.

The update rule for the inner-loop parameters is as follows:

$$\phi_i^1 = f(D_i, \phi_0) + \alpha \nabla_{\phi_0} \hat{J}(D_i, \pi_{\phi_0}),$$

where $\hat{J}(D_i, \pi_{\phi_0})$ represents an estimate of the returns for the policy π_{ϕ_0} on the task M_i, and α is the learning rate. For more complex tasks where convergence is more challenging, multiple adaptation steps may be required within the inner loop. To update the initial parameters ϕ_0 in the outer loop for better capturing the multi-task patterns, MAML computes the gradient of the returns from the updated policy with respect to the initial parameters:

$$\phi_0' = \phi_0 + \beta \nabla_\theta \sum_{M_i \sim \rho(M)} \hat{J}(D_i, \pi_{\phi_i^1}).$$

3 Methods

Most of my work focuses on the inner loop of MAML. By applying SPF, the input obtained from the state space is processed into features in the frequency domain. These features are then fed into the action policy network to take the optimal action, while also being input into the value network along with the selected actions to calculate the value and advantage at that moment. However, the MAML algorithm requires separate updates for the inner and outer loop networks. Therefore, I need to clone the network so that the cloned network shares the computation graph with the original network, enabling meta-learning.

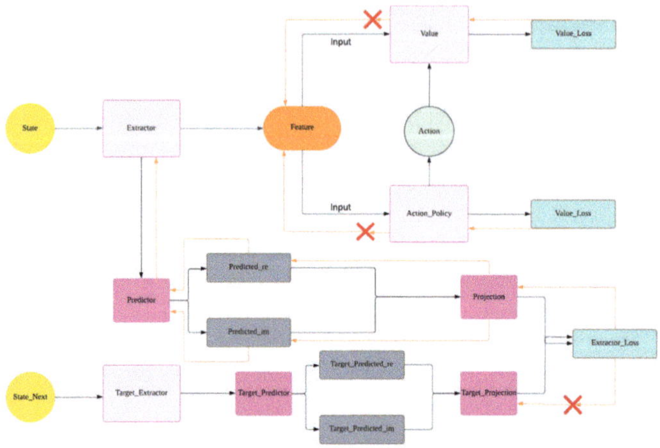

Fig. 1. The figure above shows the training update process of FMAN in the inner loop. The `Extractor` is used to extract the periodic underlying patterns in the input state, providing input to both the `Action_Policy` network and the `Value` network, which each calculate their own loss for backpropagation. `State_Next` is a state and action sampled from the data buffer, accumulated during previous training. The orange lines show how the gradients are backpropagated through the model, with gradient truncation preventing the loss from the `Action_Policy` network and the `Value` network from updating the `Extractor`, and preventing the gradient of the prediction loss from updating the `Target_extractor`.

3.1 Fourier Transform Towards State Sequence

(Ye et al., 2024) first uses Fourier transform to facilitate learning expressive representation in reinforcement learning. My work is based on the State Sequences Prediction via Fourier Transform (SPF) method proposed in their work.

For each task, I consider the standard MDP framework. The environment is denoted as $\mathcal{M} = \langle S, A, R, P, \mu, \gamma \rangle$, where S is the set of states, A is the set of actions, R is the reward function, P is the transition probability function, μ is the initial state distribution, and γ is the discount factor. I use an encoder function $\phi : S \rightarrow S$, which maps the original state $s \in S$ to a latent space $\bar{S}$. I define a latent MDP $\mathcal{M} = \langle \bar{S}, A, R, P, \mu, \gamma \rangle$, where the action space A is shared by $\mathcal{M}$ and A.

Given the current state s_t and current action a_t, I define the expectation of future state sequence as

$$[\bar{s}_t]_n = \mathbb{E}_{\pi,P}\left[s_{t+n+1} \mid s_t, a_t\right], \quad n = 0, 1, 2, \ldots \quad n = -1, -2, -3, \ldots$$

The DTFT functions can be written in a recursive form:

$$F_{\pi,P}(s_t, a_t) = \bar{S}_t + \gamma F_{\pi,P}\left[F(s_{t+1}, a_{t+1})\right].$$

The recursive relationship can be denoted as contraction mapping T, which was proved by (Ye et al., 2024). Therefore, in tabular settings, I can iteratively apply T until convergence.

Based on the recursive formula, I can obtain the prediction loss by calculating the difference between the older prediction $F_{\pi,P}(s_t, a_t)$ and the better estimate $\bar{S}_t + \gamma F_{\pi,P}\left[F(s_{t+1}, a_{t+1})\right]$. I use the prediction loss to update the parameters of the extractor and the predictor so that the encoder will output better representation that embodies more accurate information about the state sequence. I define the auxiliary prediction loss as:

$$L_{\text{pred}}(\phi, F) = d\left(\bar{S}_t + \gamma F(s_{t+1}, \pi(s_{t+1})), F(s_t, a_t)\right)$$

where $\bar{S}_t = \phi(s_t)$ means the representation of state s_t, d is an arbitrary similarity measure. I choose d as cosine similarity in experiments.

In this way, FMAN can convert the original real-valued state sequence into a complex-valued sequence to predict future states and rewards. Finally, the processed features are provided as inputs to the action policy network and value network for training.

3.2 Action Policy-Value Networks

In the FMAN architecture, I have built action policy and value networks suitable for meta-learning. Both networks utilize Fourier transform features of the input states. This approach is intended to facilitate learning in continuous action spaces and to better capture the underlying patterns behind the periodicity in the state-action space.

The action policy network is used to generate a policy that outputs a distribution of possible actions based on features. The input to the network is the Fourier transform features of the state, and it outputs the mean and standard deviation of the action distribution, represented as a Gaussian distribution. The network structure is a feedforward neural network with multiple hidden layers, each followed by a Tanh activation function. The agent samples actions from the distribution returned by this network, and the training objective is to maximize the expected return.

The value network is used to evaluate the actions taken by the action policy network, approximating the expected return for a given state. Similar to the action policy network, the value network also accepts the Fourier transform features of the state and the current action as input and outputs a scalar representing the state value. The value network consists of fully connected layers, and its goal is to minimize the temporal difference error, thereby helping to optimize the policy of the action policy network.

Both of the above networks utilize Fourier transform features as inputs, which are obtained by applying the Fourier transform function to the state sequence. This enables FMAN to capture the underlying patterns of periodicity and other frequency-based patterns inherent in the state-action trajectories. Specifically, the state sequence is transformed into a latent space, which is then used by the action policy and value networks to make more effective decisions in continuous action spaces.

By integrating Fourier transform features, FMAN is able to effectively learn from the periodic behaviors in the environment, thus improving performance in tasks that require fine-grained action decisions based on periodic state transitions.

3.3 Bi-level Optimization

FMAN's meta-parameter θ consists of the initial parameters for both the **FT** predictor and the policy network. Following the basic MAML structure, FMAN can find a much better initialization for the inner-loop network, resulting in a near-optimal inner-loop algorithm f_θ for the testing sets.

4 Experiments

In the specific experiment, I am currently using a goal-centered meta-algorithm benchmark based on Meta-World to evaluate the FMAN algorithm. In the simplest ML1 task, training is conducted using a single Meta-World task, where the "task set" for meta-training corresponds to different actions such as reaching, pushing, and pick-and-place. In each task's training process, the target object's position has different initial and goal positions, and a total of fifty tasks are set in this Meta-World setup. Meta-training is conducted at fixed initial positions to test whether the agent can apply the learned strategy to solve the problem. I evaluate the algorithm on three different task scales in Meta-World: multi-task evaluation, which involves reinforcement learning on different tasks; meta-algorithm evaluation, where the meta-learned agent solves new tasks to test if it can converge quickly. The variables in the environment include changes in the target position or object position. The target position is not included in the observation, which forces the meta-reinforcement learning algorithm to adapt to the target through trial and error. The current MAML algorithm does indeed learn and optimize its strategy throughout the training process. Success rates and returns show significant improvement in the early stages, but after reaching a certain level, performance fluctuates, and there is limited room for further improvement. The model faces challenges in further optimization and requires more samples to train when learning to tackle new tasks.

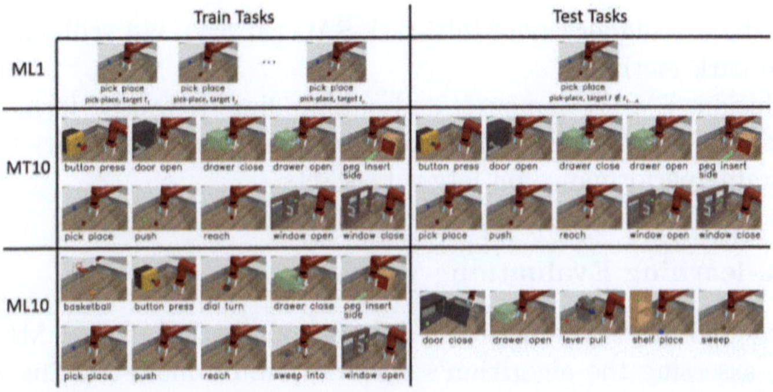

Fig. 2. The above figure visualizes three multi tasks and meta-algorithms evaluation scenarios, ranging from single-task adaptation in ML1, to training across ten different sub-tasks in MT10, and finally, to adaptation to new tasks after meta-learning in ML10. [10]

4.1 Multi-task Evaluation

In multitask evaluation [10], I compared the performance of several traditional reinforcement learning algorithms with FMAN, including multitask proximal policy optimization, multitask trust region policy optimization, multitask soft actor-critical, and task integration. These algorithms adapt to task variations in different ways when handling multi-task settings, and their success rates were evaluated on the MT10 and MT50 task sets.

Table 1. The figure shows the average maximum success rate across all tasks in the MT10 and MT50 benchmarks on three fixed seeds, with the best performance in each benchmark bolded. FMAN achieved notable advantages in multi-task evaluation across all benchmarks.

Methods	MT10	MT50
MT+FMAN	**53.5%**	**39.0%**
MT+PPO	30.5%	35.4%
MT+TRPO	31.3%	21.0%
TE	20.9%	11.8%
MT+SAC	48.3%	38.5%

On the MT10 task set, FMAN achieved a success rate of 53.5%, significantly higher than methods like PPO and TRPO, and slightly better than Multi-task SAC. This indicates that FMAN can quickly and effectively learn the periodic patterns of different tasks when handling small-scale tasks, reducing the impact of task differences on model training and optimizing its policy. However, as the number of tasks expanded to 50 (MT50), FMAN's success rate was 39.0%, slightly higher than Multi-task SAC (37.5%), but still outperforming other benchmark methods.

In conclusion, FMAN demonstrated outstanding overall performance across different test benchmarks, showing strong robustness and the ability to maintain relatively stable performance on larger task sets.

4.2 Meta-learning Evaluation

In the meta-algorithms evaluation, I focused on the ML10 and ML45 benchmarks [10], assessing the algorithm's performance on meta-training and meta-testing tasks. Compared to existing traditional reinforcement learning methods, FMAN showed advantages in many areas, especially in the meta-training phase of the ML10 task, with a success rate of 58.2%, which is much higher than MAML and PEARL. In the meta-testing tasks of ML45, FMAN achieved a success rate of 30.0%. Although this is lower than MAML and RL2, it still demonstrates strong generalization capability.

Table 2. In the meta-test evaluation phase, the figure reflects the average maximum success rate across all tasks in ML10 and ML45 for different models. The bolded parts indicate the best-performing model within the same benchmark. It is clear that the FMAN model outperforms other traditional meta-algorithms on the ML10 task set. Although its performance on the ML45 task set is somewhat underwhelming, overall, its performance surpasses that of existing traditional meta-reinforcement learning models.

Method	ML10 (Meta-Training)	ML10 (Meta-Testing)	ML45 (Meta-Training)	ML45 (Meta-Testing)
FMAN	**58.2%**	**37.5%**	49.8%	30.0%
MAML	44.4%	31.6%	40.7%	**39.9%**
RL2	56.9%	35.8%	**60.0%**	28.3%
PEARL	23.2%	13.0%	14.5%	22.0%

From the results, it is clear that FMAN outperformed MAML in the ML10 task, particularly in the meta-training phase, where the success rate was 58.2%. This indicates that FMAN is capable of quickly adapting to new tasks in meta-training and learning effective policies. Although FMAN's performance is slightly lower than MAML and RL2 on the larger-scale ML45 tasks, considering the difficulty of the MetaWorld benchmark and hyperparameter optimization issues,its success rate of 30.0% is still within a reasonable range. This suggests that FMAN has successfully extracted useful information from learning periodic patterns. By predicting long-term rewards and returns, it has had an indelible positive impact on the agent's policy formulation.

5 Discussion

This study designed FMAN (Fourier Model Agnostic Network) and evaluated its performance on multi-task learning and meta-learning tasks based on the Meta-World benchmark. The experimental results demonstrate that FMAN shows clear advantages in overall performance across multiple task sets, especially in task adaptation and efficient learning, outperforming existing traditional reinforcement learning algorithms. By introducing Fourier transforms to extract frequency-domain information, FMAN enables the agent to better understand the periodic patterns and long-term dependencies of tasks, thereby improving meta-learning performance and sample efficiency.

Upon analyzing the experimental results, I found that FMAN exhibits superior overall performance in current meta-algorithms, especially in handling continuous action spaces and periodic tasks, where it shows high learning efficiency. Although its policy converges more slowly in some complex environments, leaving room for further performance improvement, FMAN's efficiency and stability provide strong support for future research in meta-learning.

6 Conclusion

Future work will focus on enhancing FMAN's generalization ability on larger-scale tasks and further optimizing its adaptability in more complex tasks and environments. Additionally, exploring how to incorporate more prior knowledge and optimization strategies into FMAN, as well as improving model parameter initialization, will help further enhance its performance and practicality in meta-learning.

Further research can also explore how to enhance FMAN's application potential in more complex tasks and environments. Although FMAN has demonstrated excellent performance in multi-task learning and meta-learning tasks, it still faces some challenges as task diversity and environmental complexity increase. For instance, how to handle high-dimensional input data more effectively, how to maintain adaptability when faced with dynamically changing task distributions, and how to maintain efficient learning and convergence speed in large-scale tasks are important issues that future research needs to address. Incorporating strategy optimization methods from reinforcement learning, such as balancing exploration and exploitation, could also provide new directions for improving FMAN's performance. Through these explorations and optimizations, FMAN has the potential to become a more efficient, flexible, and highly generalizable meta-learning algorithm, playing a significant role in the realization of future AGI.

Acknowledgements. The authors would like to thank Peter Zhang for his valuable discussions. The authors also thank the team members at Neoscholar for their support. Zhe Hao and ZhiHeng Lai contributed equally to this work and should be considered co-first authors.

Declarations

Funding. Not applicable.

Conflict of interest. The authors declare no conflict of interest.

Ethics approval. Not applicable.

Consent to participate. Not applicable.

Consent for publication. Not applicable.

Availability of data and materials. The code for this research is available at https://github.com/gdsafdgdf/FMAN.

Authors' contributions. Zhe Hao is responsible for the model framework construction, the design and implementation of the FMAN model, the experimental parts, and the layout and writing of the paper. Zhi-Heng Lai is responsible for the design of the FMAN model extractor and the overall model design. Xuan-Yu Xue is responsible for literature research and organizing the ideas.

References

1. Volodymyr, M.: "Playing atari with deep reinforcement learning," arXiv preprint arXiv:1312.5602, 2013
2. Tobias,J., et al.: "Residual reinforcement learning for robot control," In: 2019 International Conference on Robotics and Automation (ICRA), IEEE, pp. 6023-6029 (2019)
3. Nina, M., Sergei, S., Sergei, I., Evgeny,B.: "Reinforcement learning for combinatorial optimization: a survey,"Comput. Oper. Res. vol. 134, pp. 105400, 2021
4. Jake, S., Kevin, S., Richard,Z.: "Prototypical networks for few-shot learning," In: Advances in Neural Information Processing Systems, vol. 30, 2017
5. Antreas, A., Harrison, E., Amos, S.: "How to train your MAML," arXiv preprint arXiv:1810.09502, 2018
6. Mingxuan, Y., et al.: "State sequences prediction via fourier transform for representation learning," Adv. Neural Inf. Proc. Syst. vol. 36, pp. 67565-67588, 2023
7. Eugene A F., Adam, S.: Handbook of markov decision processes: methods and applications, vol. 40, Springer Science and Business Media, 2012
8. Guy, S., David, H., Ronen I, B., Craig, B.: "An MDP-based recommender system," J. Mach. Learn. Res. vol. 6, no. 9, pp. 1265-1295, 2005
9. Jacob, B., et al.: "A survey of meta-reinforcement learning," arXiv preprint arXiv:2301.08028, 2023

10. Tianhe, Y., et al.: "Meta-world: a benchmark and evaluation for multi-task and meta reinforcement learning," In: Conference on Robot Learning, PMLR, 2020, pp. 1094-1100
11. Hao, L., Richard, S., Caiming, X.:"Taming MAML: efficient unbiased meta-reinforcement learning," In: International Conference on Machine Learning, PMLR, 2019, pp. 4061-4071
12. Harkirat, B., Baydin, S., Güneş, A., Philip HS, T.: "Alpha MAML: adaptive model-agnostic meta-learning," arXiv preprint arXiv:1905.07435, 2019

Cloud Computing: Case Studies, Adoption Models, and Theoretical Frameworks

Chiang Liang Kok(✉) and V. G. Gurupandian

The University of Newcastle, Newcastle, Australia
{chiangliang.kok,vggurupandian}@newcastle.edu.au

Abstract. Cloud computing is changing the way technology is used by providing easy access to computing resources over the internet. This article looks at real-world examples from different industries to explore how cloud computing is being used. It covers cloud service models (IaaS, PaaS, SaaS) and deployment types (public, private, hybrid), and explains how cloud systems are structured. We discuss why companies choose cloud computing – such as lower costs, flexibility, and faster innovation – as well as concerns like data security and legal compliance. The article also uses technology adoption theories (like the Technology Acceptance Model and Diffusion of Innovations) to explain how organizations decide to move to the cloud. Case studies from healthcare, retail, finance, manufacturing, and education show how cloud computing is applied in practice and what challenges are common. The article ends with strategies for successful cloud adoption and ideas for future research. The main message is that cloud computing is more than just a tech upgrade – it also requires changes in leadership, planning, and company culture.

Keywords: Cloud Computing · Cloud Adoption · Case Studies · Adoption Models · Theoretical Frameworks · Digital Transformation · IaaS · PaaS · SaaS · Technology Adoption

1 Introduction

Cloud computing has emerged as a disruptive paradigm in IT service delivery, fundamentally changing how organizations access and use computing resources [1]. This model provides on-demand remote access to a shared pool of configurable computing resources via the internet, offering unprecedented flexibility and scalability in resource management [2]. The U.S. National Institute of Standards and Technology (NIST) defines cloud computing as a model for convenient, on-demand network access to a shared pool of configurable resources that can be rapidly provisioned and released with minimal management effort [3]. In practice, cloud environments come in various forms, including public, private, and hybrid clouds, along with service models such as Infrastructure as a Service (IaaS), Platform as a Service (PaaS), Software as a Service (SaaS), and even Function as a Service (FaaS) – each providing different levels of control and abstraction to the user [4]. These options give businesses the flexibility to choose cost-effective IoT

L. Meng (Ed.): CCCN 2025, SCT, pp. 82–91, 2025.
https://doi.org/10.1007/978-3-032-03632-2_7

solutions [5] tailored to their needs and have contributed to the widespread interest in cloud adoption across sectors, and to fully realize the potential of Cloud computing, Cloud service providers have to ensure that they can be flexible in their service delivery to meet various consumer requirements, while keeping the consumers isolated from the underlying infrastructure [6]. Cloud computing architecture typically consists of two sides: the front-end (client side) and the back end (cloud provider side). The front-end includes the client devices and applications used to interact with the cloud (such as web browsers or mobile apps), while the back end encompasses the cloud servers, data storage, and service delivery infrastructure managed by the provider. Figure 1 illustrates a basic cloud computing architecture, showing how client-side interfaces connect via the Internet to back-end services and resources hosted in the cloud.

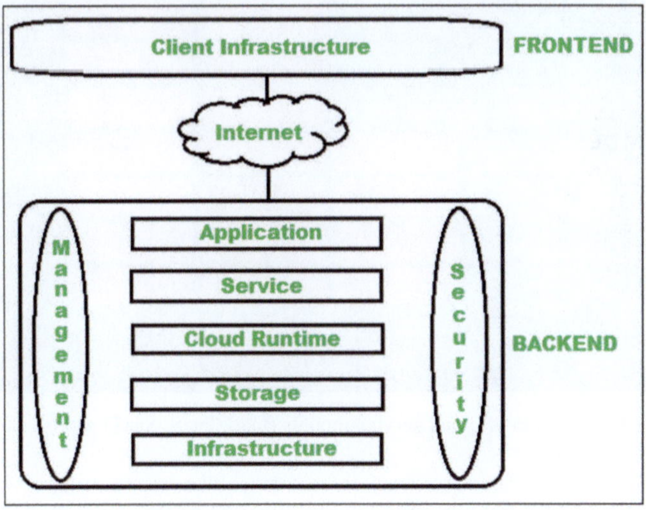

Fig. 1. Cloud computing architecture, comprising front-end client interfaces (e.g., user applications) and back-end cloud infrastructure (servers, storage, networking, management and security).

Cloud services are commonly categorized into three primary service models: Infrastructure as a Service (IaaS), Platform as a Service (PaaS), and Software as a Service (SaaS) [7].

- **IaaS** provides virtualized computing resources over the internet, such as virtual machines, storage, and networks, allowing users to manage operating systems and applications while the cloud provider manages the underlying infrastructure.
- **PaaS** offers a development and deployment platform in the cloud, supplying tools and runtime environments for developers to build, test, and deploy applications without worrying about managing the underlying hardware or OS, which are handled by the provider.
- **SaaS** delivers full-fledged software applications over the internet on a subscription basis, relieving users from installing or managing the software on local devices.

These service models can be visualized as layers of abstraction, with IaaS at the base (offering most control to the user and responsibility for the provider only at the infrastructure level) and SaaS at the top (where the provider manages almost everything, and the user simply uses the application). Figure 2 depicts the division of responsibilities between the customer and provider in on-premises IT versus IaaS, PaaS, and SaaS cloud models.

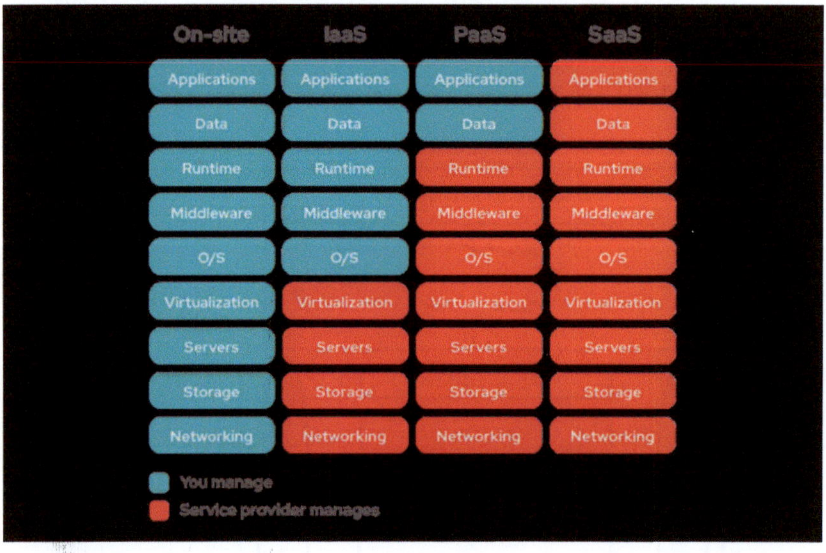

Fig. 2. Comparison of cloud service models (IaaS, PaaS, SaaS) and on-site IT.

Beyond these service models, cloud computing can be deployed in different forms: public clouds (services offered over the public internet by third-party providers to multiple customers), private clouds (infrastructure operated solely for a single organization, offering greater control and privacy), hybrid clouds (integrations of public and private cloud components allowing data and application portability), and community clouds (shared by several organizations with common concerns). This flexibility in deployment allows organizations to choose strategies that balance their needs for control, security, and cost-effectiveness. Despite its advantages, adoption of cloud computing is influenced by several factors and requires careful consideration of challenges. In the following sections, we review the drivers and barriers of cloud adoption and discuss theoretical models that researchers and practitioners use to understand and predict the adoption of cloud technologies in organizations. We then present case studies from various industries to illustrate how cloud computing is being implemented in practice, followed by an analysis of patterns and lessons learned, before concluding with recommendations for successful cloud adoption.

2 Literature Review

Cloud Computing Adoption: Drivers and Barriers

Organizations are turning to cloud computing for several important reasons. One of the biggest advantages is cost savings, as cloud services reduce the need for expensive hardware and turn large upfront costs into manageable ongoing expenses. Cloud platforms also offer great flexibility and scalability, allowing businesses to quickly adjust their computing power based on demand. This is especially helpful for growing companies or those with changing needs. Additionally, cloud services support faster innovation because developers can access tools and platforms instantly, speeding up the development and launch of new applications. Together, they emphasize how cloud computing enables a utility-like model of computing, fundamentally changing how organizations provision and use IT resources. Despite these benefits, there are challenges that can slow down cloud adoption. Security and privacy are major concerns, as storing sensitive data off-site raises fears about data breaches and losing control of important information [8]. Following data protection laws can also be more difficult when data is stored in multiple countries. Businesses also worry about service outages that can interrupt operations. Another concern is vendor lock-in, where it becomes hard or costly to switch cloud providers after committing to one. Other challenges include performance issues, like delays in applications that need real-time responses, and the difficulty of connecting cloud systems with older on-site technologies. Many organizations also face a shortage of skilled cloud professionals and must train their existing staff. Finally, company culture can impact adoption—some teams may resist the shift from traditional IT systems. For cloud adoption to succeed, companies need strong leadership, clear planning, and a willingness to embrace change.

Table 1 summarizes some common drivers and barriers of cloud computing adoption identified in the literature:

Table 1. .

Drivers	Barriers
Cost savings (pay-per-use model)	Security concerns (data breaches, privacy)
Scalability and elasticity	Compliance and regulatory issues
Faster deployment and time-to-market	Vendor lock-in risk
Accessibility and ubiquitous access	Integration with legacy systems
Innovation through advanced services	Performance (latency, downtime concerns)
Focus on core business (outsourced IT)	Lack of cloud expertise in staff
Global collaboration and mobility	Organizational resistance to change

Many of these factors have been highlighted in prior studies. For example, a survey by Rittinghouse and Ransome noted that while cost reduction and flexibility are major incentives, worries about data security and loss of control have been major impediments to cloud uptake [9]. Similarly, issues of data confidentiality and trust in the provider must be addressed for organizations to fully embrace cloud solutions [10]. Researchers often investigate these drivers and barriers through theoretical lenses, which we discuss next, to better understand how organizations decide to adopt cloud computing and under what conditions adoption is successful.

Case Studies

2.1 Communications: Cloud Adoption for Communications Networks by Alpha Mobile

Alpha Mobile, founded in 2003, creates communication systems that include corporate Short Message Service (SMS). According to one survey, the firm ranks first among the fastest growing Small and Medium Enterprises (SME) in Rio Grande do Sul, is the sixth largest corporation in the Southern Region, and ranks seventeenth in Brazil. Between 2008 and 2010, the company's net income increased by 244.8% to almost R$19 million last year. Alpha Mobile has adopted both public and private clouds, which combined account for over 40% of its infrastructure and services [11].

2.2 IT: Cloud Adoption in IT by Mainstream Renewable Power

Mainstream Renewable Power, founded in 2008, actively implemented a sustainable cloud computing approach that matched their aim of promoting renewable energy. Led by CIO John Shaw, the company focused sustainability and integration in its IT infrastructure, consulting with the Innovation Value Institute for professional advice. Mainstream improved IT maturity by conducting systematic assessments (Executive IT-CMF, SICT, and Cloud assessments), which effectively aligned business and IT strategies.

Their sustainable cloud effort lowered the IT carbon footprint, established future-proof collaborations with sustainability-focused vendors, improved energy efficiency with provider Eircom, and incorporated internal practices such as electronic processes and paper reduction. This strategy not only increased the company's sustainability but also served as a good model for others seeking to integrate environmentally friendly IT solutions [12].

2.3 Education: Cloud Computing Adoption in Zanzibar's Universities

This study explored challenges faced by two prominent Zanzibar universities—SUZA and ZU—in adopting cloud computing. A survey was conducted, involving 84 respondents, including students, staff, and lecturers, who completed questionnaires sent via email and phone. The survey responses were fully valid and carefully analyzed. The research used Partial Least Squares Structural Equation Modeling (PLS-SEM) through SmartPLS 3 to evaluate the collected data. Respondents represented both universities fairly evenly and included diverse genders and educational backgrounds. The study confirmed strong reliability and validity in its results. The statistical model demonstrated a high level of accuracy ($R^2 = 0.937$), indicating that the findings were robust and reliable.

Significant challenges identified were data security, privacy concerns, risk management, government regulations, and peer pressure. Statistical analysis confirmed these issues were significant barriers to cloud computing adoption. To overcome these barriers, the researchers recommended the adoption of the TOGAF framework [13]. This structured approach could significantly improve data security, management, and privacy. Implementing such frameworks could help universities effectively manage their cloud computing environments and achieve their academic and operational goals. Overall, the research highlighted important issues that need addressing for successful cloud computing adoption and emphasized structured, strategic solutions for universities in Zanzibar [14].

3 Theoretical Frameworks for Cloud Adoption

To analyze cloud computing adoption, it is useful to draw on established theories of technology adoption and diffusion. Several theoretical models have been applied to cloud adoption research to explain an organization's intent or decision to adopt cloud services:

- **Technology-Organization-Environment (TOE) Framework:** The TOE framework posits that three contexts influence a firm's adoption of new technology: technological factors (perceived benefits, complexity, compatibility, etc.), organizational factors (size, top management support, readiness, IT expertise), and environmental factors (industry pressure, competition, regulatory environment). In the context of cloud computing, the TOE framework has been widely used to categorize facilitators and inhibitors. For instance, within the technological context, relative advantages (improvements over existing IT) and complexity will affect adoption; within the organizational context, factors like the company's IT capability and management support are important; and in the environment, things like competitive pressure or compliance requirements play a role. Application of the TOE framework to study cloud adoption in Taiwan and found that factors such as compatibility with existing systems, firm size, and external pressure significantly influenced the decision to adopt cloud services [15].
- **Diffusion of Innovations (DOI) Theory:** Originally formulated by Everett Rogers, DOI theory explains how, why, and at what rate new technologies spread in a social system. It defines adopter categories (innovators, early adopters, early majority, late majority, laggards) and characteristics of an innovation that influence adoption, such as relative advantage, compatibility, complexity, trialability, and observability. In cloud computing, DOI has been used to understand how quickly cloud services are being adopted across industries and which characteristics of cloud innovations (e.g., ease of use, demonstrable results) drive faster uptake. Organizations that see a strong relative advantage in cloud computing and observe successful use cases in their industry are more likely to adopt it sooner [16]. Conversely, those perceiving high complexity or risks may fall into the late majority or laggard category. Figure 3 illustrates the Technology Adoption Lifecycle (adapted from Rogers' DOI theory), showing the typical distribution of adopter categories over time for a new technology.
- **Technology Acceptance Model (TAM):** TAM is a theory that specifically looks at users' acceptance of technology, focusing on two key perceptions: Perceived Usefulness (the degree to which using the technology will enhance job performance) and Perceived Ease of Use (the degree to which using the technology will be free of effort)

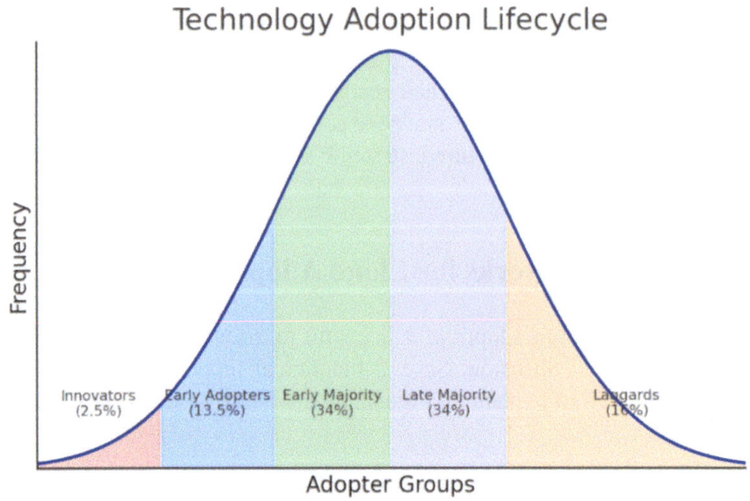

Fig. 3. Technology Adoption Lifecycle curve (based on Rogers' Diffusion of Innovations).

TAM suggests that these perceptions influence the attitude towards using the technology, which in turn affects the intention to use and actual usage. While TAM was originally developed for individual user adoption, it has been extended to organizational settings and applied to cloud services [17]. For example, if decision makers perceive cloud computing as useful for their organizational efficiency and relatively easy to implement, they are more likely to have a positive attitude and intention towards adopting it. In an enterprise context, perceived trust and security have also been studied as additional factors impacting cloud adoption, analogous to ease of use and usefulness in motivating acceptance. Figure 4 shows a schematic of the basic Technology Acceptance Model. In the cloud context, one might map "external variables" to factors like cost or vendor reputation, which influence perceived usefulness and ease of use of cloud services.

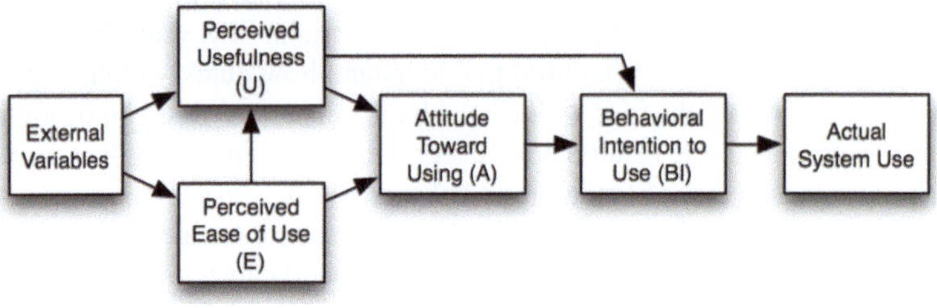

Fig. 4. Technology Acceptance Model (TAM). External variables (such as organizational and technological factors) influence users' perceived usefulness (PU) and perceived ease of use (PEOU) of technology.

- **Other Models and Frameworks:** Researchers have also integrated models to better capture cloud-specific adoption issues. For instance, some studies combine TAM with TOE (sometimes called a combined TAM-TOE model) to incorporate both individual perceptions and broader contextual factors [18].

The **Unified Theory of Acceptance and Use of Technology (UTAUT)** has been employed to examine cloud adoption, considering factors like performance expectancy, effort expectancy, social influence, and facilitating conditions. Additionally, models addressing trust and risk are particularly relevant for cloud computing. For example, an extension of TAM that includes perceived risk and trust can provide insight into how security concerns weigh against perceived benefits in the adoption decision [19]. In summary, the literature suggests that cloud computing adoption is a multifaceted decision. A combination of technological perceptions (benefits, ease, security), organizational readiness and support, and environmental pressures will determine how readily an organization moves to the cloud. The theoretical frameworks outlined above offer lenses through which we can interpret empirical examples from industry. In the next section, we describe the methodology used to gather and analyze case studies of cloud adoption, after which we will present the case studies themselves.

4 Methodology

This study uses a qualitative, multi-case approach to explore how different industries adopt cloud computing. We combined a detailed review of existing research with real-world case studies of organizations that have implemented cloud solutions. When possible, we also included insights from surveys and other available data. The aim was to identify patterns and outcomes in different settings and link them to the cloud adoption models mentioned earlier. We chose case studies from three major sectors: communications, information technology and education. These industries were selected because they have different technological needs, rules, and stages of cloud adoption, offering a well-rounded view [20]. For each sector, we picked at least one important example based on how much information was available, the scale or uniqueness of the project, and the type of cloud service or setup used (such as SaaS, IaaS, or public/private/hybrid models). Information for each case came from various sources, including company reports, industry articles, academic papers, and interviews when available. We collected consistent details for each case, like why the organization adopted cloud computing, how they did it, and what benefits or challenges they experienced. We then compared all the cases to find common themes—such as reasons for adoption (e.g., saving costs, needing flexibility), challenges faced (e.g., technical issues, training), and outcomes achieved (e.g., better performance, improved data access). We used well-known models like TOE (Technology-Organization-Environment), DOI (Diffusion of Innovations), and TAM (Technology Acceptance Model) to understand why certain decisions were made and what factors influenced success. To ensure our findings were reliable, we cross-checked data from multiple sources and used examples from different industries. This helped confirm our results and made them more applicable to a wider range of situations. By combining research with real-life examples, our method not only tells the story of each case but also connects them to larger trends in cloud adoption. While our focus is on in-depth case studies, we also refer to broader research to support our insights. The next section shares the case studies by industry, explaining how cloud computing was used and what the results were in each case.

Results Analysis and Discussion

The case studies demonstrate diverse approaches and outcomes in cloud computing adoption across different sectors: communications, IT, and education. Alpha Mobile exemplifies successful cloud integration in the communications sector, demonstrating significant growth and operational efficiency by strategically combining public and private clouds. The firm's remarkable financial growth underscores the potential economic benefits of cloud infrastructure adoption, reflecting its agility, scalability, and cost-effectiveness in competitive markets.

Mainstream Renewable Power showcases how cloud adoption aligns with strategic sustainability goals in the IT sector. The company's proactive approach, guided by rigorous maturity assessments, has substantially reduced its carbon footprint, enhanced energy efficiency, and facilitated sustainable practices internally and externally. Their strategic collaboration with vendors and cloud service providers highlights the importance of aligning cloud computing initiatives with broader organizational values and goals.

The educational sector study focusing on Zanzibar's universities identifies critical barriers to cloud adoption, including data security, privacy concerns, government regulation, risk management, and peer pressure. Despite these challenges, the high accuracy and reliability of the statistical findings ($R^2 = 0.937$) affirm the robustness of the study. The recommendation to adopt structured frameworks like TOGAF underlines the necessity of comprehensive management practices in overcoming identified barriers, suggesting that structured methodologies can significantly facilitate the successful integration of cloud technologies in educational settings.

Across all three cases, successful cloud adoption is shown to depend not only on technological factors but also on strategic alignment with organizational objectives, effective risk management, and structured implementation frameworks. The diversity in challenges and solutions across these cases underscores that cloud computing adoption strategies must be tailored specifically to organizational contexts and needs.

Conclusion

The analysis highlights that cloud computing adoption, while universally beneficial, requires sector-specific strategies tailored to unique organizational needs, challenges, and strategic goals. Alpha Mobile's significant economic gains, Mainstream Renewable Power's sustainability successes, and the structured framework recommendations for Zanzibar's universities collectively demonstrate cloud computing's versatile benefits and complexities. Effective adoption relies on aligning technological capabilities with strategic objectives, addressing security and regulatory concerns proactively, and employing structured implementation frameworks. Organizations considering cloud adoption should carefully assess their specific needs, strategically engage relevant stakeholders, and adopt comprehensive frameworks to fully harness cloud computing's potential.

References

1. Armbrust, M., et al.: A view of cloud computing. Commun. ACM **53**(4), 50–58 (2010)
2. Babar, A., Kumbhar, R.: Cloud computing adoption and migration: a systematic literature review. Comput. Ind. **121**, 103270 (2020)
3. Mell, P., Grance, T.: The NIST Definition of Cloud Computing (NIST Special Publication 800-145). National Institute of Standards and Technology
4. Rajkumar, B., Singh, M.: Cloud computing and its applications: a systematic review. J. King Saud Univ. - Comput. Inf. Sci. **32**(11), 1325–1335 (2020)
5. Kok, C.L., Ho, C.K., Lee, T.K., Loo, Z.Y., Koh, Y.Y., Chai, J.P.: A novel and low-cost cloud-enabled IoT integration for sustainable remote intravenous therapy management. Electronics **13**, 1801 (2024). https://doi.org/10.3390/electronics13101801
6. Cloud Computing Security: A Survey on Service-based Models (2021). https://doi.org/10.1016//j.cose.2021.102580
7. Beloglazov, A., Abawajy, J., Buyya, R.: Energy-aware resource allocation heuristics for efficient management of data centers for Cloud computing. Future Gener. Comput. Syst. **28**(5), 755–768 (2011). https://doi.org/10.1016/j.future.2011.04.017
8. Detecting Security Breaches in IaaS, PaaS, and SaaS Networks for Enhanced Cloud Application Protection (2023). https://www.researchgate.net/publication/390399373_Detecting_Security_Breaches_in_IaaS_PaaS_and_SaaS_Networks_for_Enhanced_Cloud_Application_Protection
9. Rittinghouse, J.W., Ransome, J.F.: Cloud Computing: Implementation, Management, and Security. CRC Press (2010)
10. Hashem, I.A.T., Yaqoob, I., Anuar, N.B., Mokhtar, S., Gani, A., Khan, S.U.: The rise of "big data" on cloud computing: review and open research issues. Inf. Syst. **47**, 98–115 (2015)
11. Sobragi, C.G., Maçada, A.C.G., Oliveira, M.: Cloud computing adoption: a multiple case study (2025). Ufrgs.br. 1807-054X
12. Conway, G., Curry, E., Donnellan, B.: Cloud computing adoption: an SME case study. In: 17th Annual Irish Academy of Management Conference, p. 1 (2014). https://mural.maynoothuniversity.ie/id/eprint/10883/1/BD-Cloud-computing-2014.pdf
13. Design Enterprise Architecture Using TOGAF-ADM Framework to Support Academic Information Systems. Nusa Nipa University (2023). https://doi.org/10.1109/ICE-SMARTECH59237.2023.10461950
14. Juma, M.K., Tjahyanto, A.: Challenges of cloud computing adoption model for higher education level in Zanzibar (the case study of SUZA and ZU). Proc. Comput. Sci. **161**, 1046–1054 (2019). https://doi.org/10.1016/j.procs.2019.11.215
15. Low, C., Chen, Y., Wu, M.: Understanding the determinants of cloud computing adoption. Ind. Manag. Data Syst. **111**(7), 1006–1023 (2011)
16. Rogers, E.M.: Diffusion of Innovations, 5th edn. Free Press, New York (2003)
17. Paula, Michel, B., Chiappetta, J., Mariano, E.B., Beatriz, A.: Management theory and big data literature: from a review to a research agenda. Int. J. Inf. Manag. **43**, 112–129 (2018). https://doi.org/10.1016/j.ijinfomgt.2018.07.005
18. Bryan, J.D., Zuva, T.: A review on TAM and TOE framework progression and how these models integrate. Adv. Sci. Technol. Eng. Syst. J. **6**(3), 137–145 (2021). https://doi.org/10.25046/aj060316
19. Venkatesh, V., Thong, J.Y.L., Xu, X.: Unified Theory of Acceptance and Use of Technology: A Synthesis and the Road Ahead, May 2016. Ssrn.com. https://papers.ssrn.com/sol3/papers.cfm?abstract_id=2800121
20. Shrivastava, U., Sabharwal, S., Gupta, M.: Addressing cloud computing challenges – a review. In: Proceedings of the 2018 International Conference on Advances in Computing, Communications and Informatics (ICACCI), pp. 2374–2380. IEEE (2018)

Optimization Study of Aircraft Position Allocation Based on Improved ECA Algorithm

Jabin Feng[1(✉)], Chunle Yang[2], Tiance Yin[2], Xuezhi Xu[3], Yilei Chen[2], and Fang Zhou[4]

[1] College of Aviation Engineering, Civil Aviation Flight University of China, Guanghan, China
993343176@qq.com

[2] College of Airport, Civil Aviation Flight University of China, Guanghan, China

[3] CAAC Key Laboratory of Flight Techniques and Flight Safety, Civil Aviation Flight University of China, Guanghan, China
xuezhixu@cafuc.edu.cn

[4] Guanghan Branch, Civil Aviation Flight University of China, Guanghan, China

Abstract. To ensure the rational allocation and efficient utilization of parking space resources at civil airports, we establish a multi-objective optimization model aimed at minimizing flight delay costs, passenger walking distances, and idle parking space time. This model is developed by analyzing the fundamental safety operation rules governing parking space allocation while considering constraints such as aircraft delay conditions and the availability of bridge parking spaces during flight delays across various time slots. We propose an improved ECA algorithm that integrates the original ECA algorithm with an elite inverse strategy, enhancing its convergence speed, preventing early convergence, and increasing search diversity, ultimately leading to improved solution quality. The experimental results show that the improved ECA algorithm can optimize the flight delay time and the bridge call rate compared with the original scheme, specifically, the ECA algorithm incorporating the elite reverse strategy can improve the bridge call rate by 3% and reduce the bridge call rate by 72.16 min, and also shows faster convergence speed and stronger global optimization ability, which improves the efficiency of the air transportation, and has great significance for the management of airports and the allocation of the parking space resources. It has practical significance for airport management and parking space resource allocation.

Keywords: parking space allocation · ECA algorithm · multi-objective optimization

1 Introduction

Airport apron resource allocation stands as a pivotal determinant of operational efficiency and passenger experience in air transportation hubs. Scholars have extensively explored optimization strategies to address the dual challenges of meeting spatiotemporal constraints in flight scheduling and mitigating the impacts of escalating flight delays.

L. Meng (Ed.): CCCN 2025, SCT, pp. 92–105, 2025.
https://doi.org/10.1007/978-3-032-03632-2_8

Notable contributions include Zhang et al.'s (2023) multi-objective optimization framework, which integrates flight conflict minimization, boarding bridge utilization maximization, and taxiing fuel consumption reduction through a cosine variance-adaptive grouping cuckoo search algorithm (CG-TSA) [1]. This hybrid approach demonstrated superior convergence and robustness in computational experiments. Chen et al. (2022) advanced the field by constructing a maintenance, repair, and overhaul (MRO) apron allocation model using mathematical programming and network flow techniques. Their empirical validation, grounded in real-world service provider data, validated the model's utility as a practical planning tool [2]. Concurrently, Du et al. (2024) addressed the Robust Gate Assignment Problem (RGAP) by proposing a column generation-based heuristic (CDBH) that incorporates buffer time allocation and conflict quantification metrics, enhancing resilience against operational disruptions [3]. From a strategic planning perspective, Chun-Ying Chen (2023) developed a spatiotemporal network flow model for apron allocation, revealing that planners can achieve significant efficiency gains through scenario based optimization [4]. Similarly, Ningning Zhao (2023) formulated a multi-objective preemptive slot allocation model prioritizing passenger travel distance, airline cost minimization, and slot utilization efficiency, validated via simulated annealing algorithms [5].Despite these advancements, existing studies predominantly rely on conventional metaheuristics (e.g., genetic algorithms, particle swarm optimization) or exact methods (e.g., integer programming), with limited exploration of bio-inspired algorithms in dynamic delay scenarios.

Notably, the Empire Competition Algorithm (ECA)-a population intelligence paradigm mimicking historical imperial competition mechanisms—offers unique advantages in balancing exploration and exploitation for complex optimization problems [6]. However, its application in aviation resource allocation remains uncharted. Building upon this gap, this study introduces an elite-inverse strategy-enhanced ECA to solve the apron allocation model under flight delay conditions. Experimental results demonstrate the algorithm's efficacy in achieving high-quality solutions with reduced computational overhead, offering a novel paradigm for enhancing airport operational resilience.

The approach is the phase in which an airplane descends from its cruising altitude and prepares to land. For the approach phase, the aircraft is exposed to a number of hazards. Visibility, weather conditions, airport runway conditions, etc., are the main factors affecting the approach and landing of an airplane, and any problem in any of these factors will pose a threat to the airplane. Therefore, it is necessary to add risk identification on the basis of trajectory anomaly detection. It is needed to calculate the distance to the cluster center of the abnormal trajectories that have been filtered out by the clustering method, and to determine the risk threshold by the size of the distance [2].

Risk identification and assessment refers to the analytical process by which safety managers determine the impact on system safety in terms of the likelihood and severity of risk occurrence. For the terminal area flight approach phase, once the risk is transformed into danger, the consequences are disastrous, so the terminal area approach risk assessment should focus on studying the possibility of risk occurrence. According to the judgment of the severity of the occurrence of risk, the risk can be classified as low risk, medium risk, high risk, etc.

Based on the above problems about cluster analysis and risk identification, we took the terminal area approach trajectories as the research object, constructed the clustering and identification algorithms based on the trajectory similarity, and experimented with the problems. The main contributions of this paper include:

- We use a trajectory clustering method based on Dynamic Time Wrapping (DTW) computation to support the analysis of large-scale as well as high-density trajectory data for the increasingly complex flow of civil air traffic.
- We introduce an abnormality detection method to deal with abnormal trajectories in flight data, and devise a hierarchical risk-based statistical method to compute risk values for different abnormal trajectories. Subsequently, our method is able to recognize abnormalities in flight trajectories and perform risk assessment.
- Experiments conducted on the dataset demonstrate the effectiveness and efficiency of DTW. The size of the distance from the abnormal trajectory to the clustering center point was successfully used to set different risk levels, and finally the number of higher-risk abnormal trajectories was derived. It provides a reference for civil aviation flight safety monitoring.

The rest of the paper is organized as follows: Sect. 2 presents the background knowledge and problem description related to flight trajectory and algorithmic modeling. Section 3 provides a detailed description of anomaly identification and risk assessment. Section 4 describes the experiments and analyzes the results. Finally, Sect. 5 summarizes the work of this paper.

2 Problem Statement

This study constructs a parking space optimization model under time-of-day flight delay conditions, targeting three interrelated objectives: mitigating economic losses from flight delays through optimized resource allocation, reducing passenger walking distances via proximity-based slot assignment to enhance traveler satisfaction, and minimizing aircraft idling time to improve apron utilization efficiency. By synergistically addressing these dimensions, the model resolves the inherent trade-offs between operational cost-efficiency, passenger-centric services, and infrastructure productivity. Specifically, it aims to curtail financial burdens on airports and airlines caused by prolonged delays, streamline ground handling processes through spatially intelligent slot assignments, and maximize resource throughput by dynamically reallocating underutilized parking spaces. Collectively, these optimizations establish a resilient framework for enhancing airport operational resilience while balancing competing demands in high-density air traffic scenarios.

2.1 Hypothetical Condition

a) All flights are allocated based on the first-come, first-served principle, without employing any combined allocation strategies.
b) It is assumed that different types of flights (e.g., large aircraft vs. small aircraft) have uniform parking space demands, or are simplified using a uniform model that disregards variations in requirements due to aircraft size.

c) To reduce model complexity, it is assumed that no flight changes or transfers are required; all flights arrive directly and remain at their assigned parking spaces throughout their duration.
d) The occupancy time of a parking space is defined as the cumulative stopping time of all flights assigned to that space, with the total available time for all spaces assumed to be 24 h.
e) The model assumes that 80% of airport bridge capacity is required for operations.

2.2 Model Construction

1) *Model Parameter Definition*
a) *L:* Parking space assembly, j: Aircraft Position Number.
b) *N*: Flight assembly, *i*: Flight Number.
c) u_j: Total available time at the airfield.
d) t_i: The time of occupancy of an aircraft slot, i.e. the actual length of time the slot is in use.
e) d_i:the delay time of the flight.
f) D_{ik}: The average delayed operating cost of the flight, which is expressed as light, medium and heavy turbulence according to the strength of the turbulence of the aircraft.
g) w_{ij}: The walking distance from the security checkpoint to the boarding gate for passengers when the flight is parked at the aircraft stand.
h) R^*: The ratio requirement of the airport's docking at the bridge.

The decision variables are:

Table 1. Average delayed flight operating costs

models	representative symbol	Type of wake	Average delayed operating costs ($/h)
Light machine	***L***	Lightweight Tail Stream	208
medium-sized machine	***M***	Medium-sized wake	2916
Heavy duty	***H***	Heavy Duty Tail Stream	4167

X_{ij}: Flight assigned to a parking space; if assigned, then
X_{ij} =1, otherwise X_{ij} =0。
B_j: Whether the slot is a bridge-dependent slot, if so, then $B_j = 1$, otherwise B_j=0。

2) *Objective function*

In order to establish the optimization model of parking space allocation under flight delay, the conditions of minimum impact of flight delay, shortest walking distance of passengers and maximum utilization of parking space need to be met. Therefore, the model is obtained as follows:

a) Minimizing the cost of flight delays

$$Minf_1 = \sum_{N}^{i=1} D_{ik} \cdot d_i \tag{1}$$

In the formula, the average delayed operating costs of flights D_{ik} are shown in Table 1 below.

b) Minimize walking distances for passengers

$$Minf_2 = \sum_{N}^{i=1} \sum_{L}^{j=1} X_{ij} \cdot W_{ij} \tag{2}$$

c) Maximize parking space utilization

$$Minf_3 = \sum_{L}^{j=1} (u_j - t_j) \tag{3}$$

Formula,

$$t_j = \sum_{N}^{i=1} (t_i^d - t_i^a) \cdot X_{ij} \tag{4}$$

d) Enhancement of the flight dependency ratio

$$\sum_{N}^{i=1} \sum_{L}^{j=1} (1 - B_j) \cdot X_{ij} \tag{5}$$

e) Overall objective function

$$MinF = c_1 \sum_{N}^{i=1} D_i \cdot d_i + c_2 \sum_{N}^{i=1} \sum_{L}^{j=1} X_{ij} \cdot W_{ij} + c_3 \sum_{L}^{j=1} (u_j - t_j) \tag{6}$$

Among the. c_1 c_2 c_3 is a weighting factor to balance the objectives.

3) *Restrictive condition*

a) *Each aircraft must be assigned to one slot*

$$\sum_{L}^{j=1} X_{ij} = 1, \forall i \in N, \forall j \in L \tag{7}$$

b) *Allocation of up to one aircraft per slot*

$$\sum_{N}^{i=1} X_{ij} \leq 1, \forall i \in N, \forall j \in L \tag{8}$$

c) *Flight delay time constraints*

$$d_i \geq \Delta T_i, \forall i \in N \tag{9}$$

d) *Constraints on the number of machine positions against the bridge*

$$\frac{\sum_{N}^{i=1} \sum_{L}^{j=1} B_j \cdot X_{ij}}{N} \geq R^* \tag{10}$$

2.3 Model Solution

4) *Empire Competition Algorithm.*

Empire Competition Algorithm (ECA) is a heuristic optimization algorithm based on competition between empires in nature. It is inspired by the process of empire formation and decline and simulates the competitive behavior among empires. ECA was proposed by Iranian scientists Esmat Rashedi, Hossein Nizamabad-pour and Siamak Tabatabai in 2007. The core idea of ECA is to simulate the competitive behavior between multiple empires, each of which consists of a number of "individuals" that represent a solution. Among the empires, individuals constantly compete with each other to improve the quality of the solution through the process of survival struggle, resource allocation, evolution, and so on, and finally make an empire stand out and obtain the optimal solution. The whole process of evolution is similar to the law of survival competition and evolution of species in nature.

a) Initialization

State creation: randomly generate a certain number of states, each representing a candidate solution in the problem space.

Empire Formation: Based on the fitness (quality) of the countries (solutions), stronger countries (high quality solutions) are used as empires, and weaker countries are assigned to these empires as colonies.

b) *Calculate the degree of adaptation.*

Each individual is evaluated and its fitness value is calculated. The fitness $F(X_i^j)$ is the value of the objective function of the problem and can be designed according to the specific problem. Assuming that the objective function is $f(X)$, the fitness is calculated as:

$$F(X_i^j) = f(X_i^j) \tag{11}$$

where. X_i^j is the solution of the i individual in the j empire, denoted as a vector, usually the decision variable of the problem. $f(X_i^j)$ is the value of the objective function, usually the objective function of the minimization or maximization problem.

c) *Imperial Competition.*

Internal Competition: Within each empire, individuals engage in competition where less adaptable individuals are eliminated, while more adaptable ones acquire greater resources. Each generation involves a selection process where the fittest individual is chosen to survive based on fitness evaluation.

For the i empire, the most adapted individual is selected as the "leader" individual:

$$X_i^* = arg \max_{j=1,2,\ldots,N} F(X_i^j) \tag{12}$$

External competition: Empires compete based on their overall fitness, with higher-fitness empires gaining advantages in resource allocation. This enables them to support larger populations and exert greater influence over the system.

The fitness of each empire can be assessed by the fitness of its optimal individual:

$$F(E_i) = F(X_i^j) \tag{13}$$

Formula E_i is the first empire

d) *Resource allocation.*

Resources are allocated based on the empire's level of adaptation, with empires with higher levels of adaptation receiving more resources for improving their individuals. Resources $R(E_i)$ are proportional to the empire's fitness $F(E_i)$, and can specifically be allocated by normalization:

$$R(E_i) = \frac{F(E_i)}{\sum_{N}^{i=1} F(E_i)} \tag{14}$$

e) Individual updating and generating new solutions

After competition and resource allocation, individuals in each empire need to be renewed. Renewal can take the form of mutations, crossovers, random wandering, etc. Assuming that new individuals $X_i^{j'}$ are generated through mutation operations, the updating process can be expressed as follows:

$$X_i^{j'} = X_i^j + \sigma X_i^j \tag{15}$$

where σX_i^j is the change brought about by the mutation operation, and the specific method can be designed based on the problem characteristics.

If the crossover operation is used, two individuals X_i^{j1} and X_i^{j2} undergo crossover to generate a new individual:

$$X_i^{j'} = C(X_i^{j1}, X_i^{j2}) \tag{16}$$

where C is the crossover operation and $X_i^{j'}$ is the new solution generated.

f) *Empire declines and extinction*

As the iteration proceeds, the fitness of some empires gradually decreases, and they may move towards decline or even extinction. If the fitness value of an empire falls below a certain threshold value of θ, the empire will die out and be replaced with the best individuals from other empires or a new empire:

If

$$F(E_i) < \theta \tag{17}$$

Then empire E_i declines, while a new empire E_{new} is generated based on the optimal individuals or by migrating individuals from other empires.

g) *Iterative update and stopping criterion.*

The above steps are repeated until the stopping criterion is satisfied. The stopping criterion can be that the maximum number of iterations is reached T_{max} or the change in the objective function is less than a set threshold δ:

If

$$t \geq T_{max} \tag{18}$$

or

$$\max_i \mid F(E_i) - F(E_i^{(t-1)}) \mid < \sigma, stop \tag{19}$$

where t is the current number of iterations and $_\sigma$ is the stopping accuracy threshold.

h) *Output the optimal solution.*

Finally, the individual X_{best} with the optimal fitness is output, and the corresponding fitness value of this individual is the optimal solution to the problem:

$$X_{best} = arg\ \max_i F(X_i^*) \tag{20}$$

5) *Optimizing the ECA algorithm*

The Empire Competition Algorithm (ECA) effectively avoids local optima by simulating competition and cooperation among multiple empires, demonstrating strong global search capabilities. It is well-suited for multi-objective optimization problems, balancing and exploring different objectives through the rational allocation of resources and co-evolution of empires, thereby overcoming the limitations of single-objective approaches. However, ECA may converge prematurely or become trapped in local optima, especially in complex solution spaces or with irregular objective functions, if it fails to maintain sufficient population diversity during the search process.

To address these limitations, we enhance the ECA by integrating an elite reverse strategy. This strategy combines elite selection with reverse learning to improve population quality and diversity. The core idea is to generate reverse solutions for elite individuals, compare their fitness with the original solutions, and selectively retain the better individuals for the next generation. This approach optimizes the initial population quality, accelerates convergence, avoids early convergence, enhances search diversity, and ultimately improves solution quality.

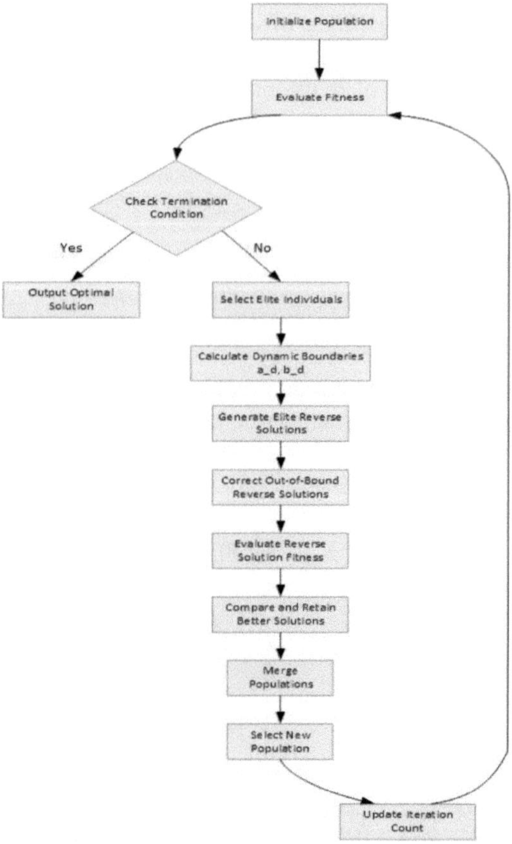

Fig. 1. Flowchart of elite reverse strategy

a) Elite Reverse Strategy

The elite reverse strategy is an optimization algorithm improvement method that combines elite strategy with reverse learning to improve population diversity, avoid local optimality, and accelerate convergence. The core idea is to generate reverse solutions for elite individuals, compare their fitness with the original solutions, and selectively retain the better individuals for the next generation. Figure 1 illustrates the flow chart of the elite reverse strategy.

Step 1 $P = \{x_i\}_{i=1}^{n}$ Population initialization and fitness assessment.

Generate an initial population1, where each individual $x_i = (x_{i1}, x_{i2}, ..., x_{iD})$ is a D-dimensional solution vector, and the initial values of each dimension are usually randomly generated within a fixed boundary. And calculate the fitness value $f(x_i)$ of each individual for subsequent screening and comparison.

Step 2 Elite Individual Selection

The population is ranked according to the fitness value, and the top of individuals are selected as the elite group $E = \{N_j\}_{j=1}^{m}$, where $m = p\% \cdot n$. The elite individuals usually represent the high-quality solutions in the current iteration.

Step 3 Dynamic boundary calculation

To avoid the fixed boundary limiting the search range, the dynamic boundary of the current population (or elite group) is calculated for each dimension d: $a_d = \min_{1 \le i \le n} x_{id}$, $b_d = \max_{1 \le i \le n} x_{id}$. Where the dynamic boundary adjusts adaptively with the population distribution and shrinks gradually to focus the optimal region.

Step 4 Inverse solution generation

For each elite individual N_j, its inverse solution $N_j^* = \left(N_{j1}^*, N_{j2}^*, ..., N_{jD}^*\right)$ is generated, and the formula for the inverse solution is further divided into two categories depending on the strategy:

Standard reverse solution $N_{jd}^* = a_d + b_d - N_{jd}$, which is expressed as directly subtracting the original solution component from the sum of the dynamic boundaries.

Dynamic inverse solution $N_{jd}^* = k \cdot (a_d + b_d) - N_{jd}$, where is a dynamically adjusted scaling factor to be used to increase randomness.

Step 5 Inverse Solution Evaluation and Selection

Compare the fitness values N_j and N_j^* of the original solution $f(N_j)$ and the reverse solution $f(N_j^*)$ and retain the better solution into the next generation population. If $f(N_j^*) > f(N_j)$, replace N_j with N_j^*, otherwise retain the original solution.

Step 6 Dynamic search space update

Merge the elite reverse solution with the original population to form an extended search space $\{x_i\} \cup \{N_j^*\}$, and filter the high-quality solutions through fitness evaluation to drive the population to converge to the global optimal region.

6) *Algorithmic step.*

The steps of the improved empire competition algorithm are shown in Fig. 2.

Step 1 Determine the appropriate algorithm parameters, initialize the populations (empires and colonies) using an elite inverse strategy to generate the initial countries and their inverse solutions, and merge the initial populations and assign empires and colonies according to each country's fitness ordering.

Step 2 Each empire directs its colonies to move towards itself through an assimilation operation.

Step 3 Generate a new solution X_i^j randomly by some colonial revolutions and generate a reverse solution X_i^{j*} by using the elite reverse strategy again after the revolutions, calculate and compare the fitness of the two and choose the one with high fitness as the optimal solution in this phase.

Step 4 Verify the relative position of each colony with respect to its imperialist country. If a colony exhibits superior adaptability compared to its affiliated imperialist country or region, the two entities undergo a role reversal, wherein the colony ascends to the status of a new imperialist country. This triggers an imperialist competition, wherein the newly elevated country seizes a colony from the weakest empire and redistributes it to the strongest empire. Empires that lose all their colonies are subsequently eliminated from the system.

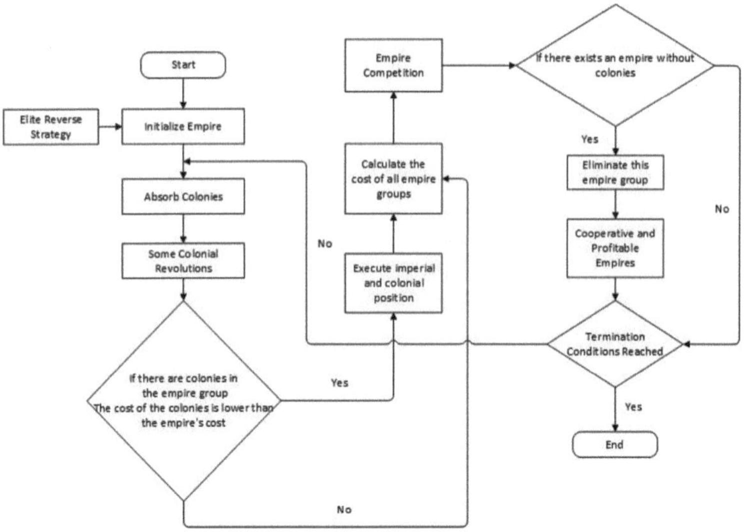

Fig. 2. Flowchart of the improved algorithm

Step 5 Repeat steps 2 to 4 until the stopping condition is satisfied (e.g., only one empire left, or the maximum number of iterations is reached), and output the optimal solution X_{best}.

3 Empirical Research

3.1 Experimental Setup

The experiments conducted in the paper are carried out in python environment, and the experimental data are the real flight operation data of a normal day at an airport in Southwest China, and the flight schedules (partial) are shown in Table 2. In terms of the aircraft position setting, this airport has a total of 125 parking spaces, which contains 91 near aircraft positions and 34 far aircraft positions. The experimental parameters are configured as 200 flights, 125 parking spaces, 100 empires, 100 maximum iterations, 0.5 probability of variation, and 0.3 elite ratio. For the determination of the experimental parameters, the one-way analysis of variance (OAT) is used to conduct a local sensitivity analysis on the probability of variance k and the elite ratio *p*. The values of the rest of the experimental parameters are fixed by varying the probability of variance *k* and the elite ratio *p*. The values of the experimental parameters are repeated many times. By varying the values of variation probability *k* and elite ratio *p* and fixing the values of the rest of the experimental parameters, repeated experiments were carried out to determine the values of variation probability *k* and elite ratio *p* in the case of the optimal allocation of machine positions.

3.2 Experimental Results and Analysis

Through the experimental analysis, the flight delay time under the improved ECA algorithm was calculated as 828.17 min, while the original scheme recorded a delay of 900.33 min. The optimization achieved a reduction of 72.16 min in flight delays. Under the improved algorithm, the flight bridging rate reached 89.0% (as illustrated in Fig. 3). Compared to the original airport's actual scheme (86.0%), this demonstrates that the enhanced ECA algorithm effectively improves the airport's bridge rate by 3.0%, with a more rational distribution of slot utilization compared to the original plan.

To visually compare the gate allocation performance of the original scheme and the optimized ECA algorithm, a Python-based simulation experiment was conducted. The final gate allocation results are summarized in Table 3, and the iterative convergence curve of the improved ECA algorithm is presented in Fig. 4.

Figure 4 illustrates that the fitness value of the improved ECA algorithm exhibits a gradual decline during iterations, indicating effective optimization of gate allocation. As the number of iterations increases, the fitness value stabilizes at a lower level of 1.89. This outcome confirms that the ECA algorithm successfully avoids local optima by integrating an elite reverse strategy in the initial population stage and a revolution mechanism in later stages, thereby identifying a globally optimal gate allocation solution.

In conclusion, compared to the original scheme, the improved ECA algorithm demonstrates significant advancements in enhancing flight bridging rates and reducing delays for flight delay mitigation scenarios. Therefore, this enhanced algorithm presents a robust solution for addressing airport gate allocation challenges.

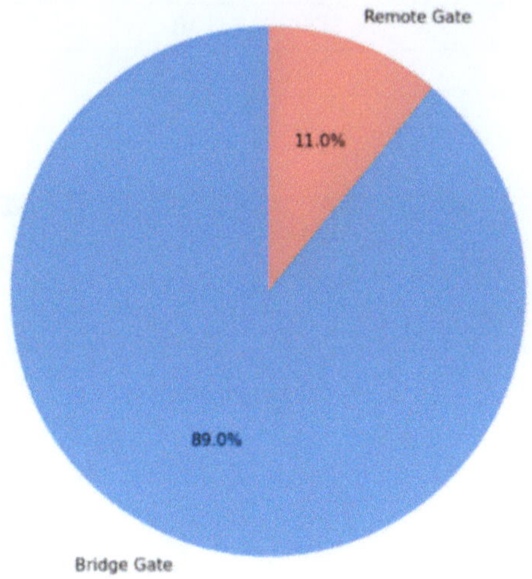

Fig. 3. Bridge Usage Rate

Table 2. Flight schedules (partial)

Flight ID	TAD	TAP	Type of wake
B1552	5:40	6:30:	M
B308S	7:55	9:55	M
B20AE	8:05	10:15	M
B8018	9:10	11:15	M
……	……	……	……
B206N	21:05	23:10	M
B8553	21:20	23:35	M
B1876	21:50	23:55	M

Table 3. Comparison of original and improved ECA algorithm allocation schemes (partial)

Flight ID	Results of the original allocation program	Improved ECA algorithm allocation results
1	12	53
2	78	40
3	98	13
……	……	……
199	67	10
200	71	2

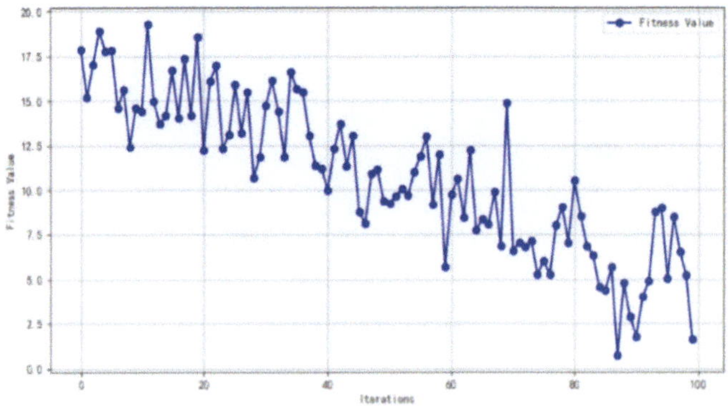

Fig. 4. Iteration graph

4 Conclusion

This study investigates the parking space allocation problem under flight delay conditions through an enhanced Evolutionary Computation Algorithm (ECA) integrating an elite inverse strategy. Drawing on empirical data from a typical day's operations at a major Southwest China airport, we establish a mathematical model and conduct comprehensive case validation. Comparative analysis between the original allocation scheme and our experimental results demonstrates the improved algorithm's enhanced reliability and operational efficiency gains, particularly in optimizing airport resource utilization.

Notwithstanding these promising results, several areas merit deeper exploration for practical implementation. Future research directions should focus on: (1) Refining the assimilation mechanism within the ECA framework; (2) Optimizing parameter configuration through systematic sensitivity analysis; (3) Developing adaptive fitness functions for diverse operational scenarios; and (4) Conducting large-scale validation using multi-source operational data spanning varied temporal and meteorological conditions.

References

1. Zhong, H., Zhang, C., Yao, X.: Membership-based aircraft parking stand allocation system with time window constraints: an event–based time–space separated algorithm. Expert. Syst. **40**(10), e134310 (2023)
2. Chen, C.Y., Chan, I.H.: The stand allocation model for aircraft MRO service provider. Transp. Plann. Technol. **47**(5), 788–808 (2024)
3. Du, W., Zhu, S., Tong, L.: Robust gate assignment to minimise aircraft conflicts. Transportmetrica B: Transp. Dyn. **11**(1), 2185497 (2023)
4. Zhang, Y., He, Q., Yang, L.: An improved tunicate swarm algorithm for solving the multiobjective optimisation problem of airport gate assignments. Appl. Sci. **12**(16), 8203 (2022)
5. Zhao, N., Duan, M.: Research on airport multi-objective optimization of stand allocation based on simulated annealing algorithm. Math. Biosci. Eng. **18**(6), 8314–8330 (2021)
6. Atashpaz-Gargari, E., Lucas, C.:Imperialist competitive algorithm: an algorithm for optimization inspired by imperialistic competition. In: 2007 IEEE Congress on Evolutionary Computation, pp. 4661–4667. IEEE (2007)

Computer Network and Information Security

Optimizing Intrusion Detection Systems with Q-Learning Techniques

Cao Tien Thanh(✉) and Do Phi Hung

Department of Information Technology, University of Foreign Languages and Information Technology, Ho Chi Minh City, Vietnam
{thanhct,hungdp}@huflit.edu.vn

Abstract. With the increasing prevalence of cyber attacks, it is crucial to develop effective strategies for defending against these threats. In this paper, we investigate the use of Q-learning, a popular reinforcement learning algorithm, for enhancing network security. We explore the development of optimal strategies for defending against cyber attacks by defining the state, action, and reward, and applying Q-learning to learn an optimal policy. Our experiments show that Q-learning can effectively learn an optimal strategy for network defense, and can outperform traditional rule-based approaches in certain scenarios. We also discuss the challenges and limitations of using Q-learning for network security, and suggest future research directions.

Keywords: Cyber attacks · Q-learning · Security · IDS · Machine learning

1 Introduction

Cyber attacks are becoming increasingly sophisticated and frequent, posing a significant threat to organizations and individuals alike. Traditional rule-based approaches to network security, such as firewalls and intrusion detection systems, are becoming less effective in defending against these threats. As a result, there is a growing interest in developing adaptive and intelligent approaches to network security.

Reinforcement learning (RL), a subfield of machine learning, is a promising approach to developing adaptive strategies for network security. Reinforcement learning algorithms learn to optimize a policy based on trial-and-error interactions with the environment. Q-learning is a popular reinforcement learning algorithm that has been successfully applied to a wide range of applications, including robotics, gaming, and finance.

In this paper, we investigate the use of Q-learning for enhancing network security. We explore the development of an optimal strategy for defending against cyber attacks by defining the state, action, and reward, and applying Q-learning to learn an optimal policy which can be applied to build IDS agent that detects the anomaly in network packets. These anomalies can be categorized into various attack types based on the network configuration. Our experiments show that Q-learning can effectively learn an optimal strategy for network defense, and can outperform traditional rule-based approaches in certain scenarios.

L. Meng (Ed.): CCCN 2025, SCT, pp. 109–126, 2025.
https://doi.org/10.1007/978-3-032-03632-2_9

As the digital landscape continues to evolve, this investigation stands as a critical step towards understanding and harnessing the potential of cutting-edge technologies in safeguarding our networks against an increasingly sophisticated array of cyber risks. Numerous studies have ventured into the integration of reinforcement learning, specifically Q-learning, within the realm of cybersecurity, aiming to revolutionize defense strategies. For instance, Hu Z et al. [1,2] likely delves into adaptive defense mechanisms, showcasing how reinforcement learning techniques dynamically respond to emerging threats. The potential advantages might include adaptability to evolving attack patterns, although challenges like scalability and real-time responsiveness may be evident. The paper titled "Enabling intrusion detection systems with dueling double deep Q -learning" by Badr Y [3] is poised to explore the design of intrusion detection systems employing Q-learning algorithms, potentially highlighting their efficiency in identifying novel threats while discussing limitations such as computational complexity and adaptability to new attack vectors.

Author Khoei et al. [4] likely investigates the synergy between deep reinforcement learning and Q-learning, presenting advancements in autonomous defense mechanisms. Here, benefits like enhanced decision-making and self-learning capabilities might be outlined, along with potential constraints such as data requirements for deep models. Moreover, Lopez et al. [5] is anticipated to introduce an innovative Q-learning-based methodology for cybersecurity, potentially offering advantages like adaptability to diverse threat landscapes but also highlighting challenges in implementation or scalability. The paper titled "CPSS LR-DDoS Detection and Defense in Edge Computing Utilizing DCNN Q-Learning" by author [6] could center on the utilization of Q-learning for identifying and countering Distributed Denial of Service (DDoS) attacks. It might highlight benefits like rapid adaptation to evolving attack patterns, enabling more responsive defense mechanisms. However, these approaches could face challenges due to the necessity for extensive and frequently updated datasets, potentially hindering their practical implementation. Overall, these hypothetical papers strive to underscore Q-learning's potential in bolstering cybersecurity while recognizing the intricacies and challenges inherent in its practical use. In summary, this paper presents the following contributions:

- Introducing an image embedding scheme, the network traffic data is transformed into time-series and subsequently into image data. This enables processing by Convolutional Neural Networks (CNN) and a 1D-CNN-based architecture of the Reinforcement Learning (RL) agent, addressing intrusion detection challenges. The dataset for packet-level IDS includes flow information to incorporate temporal details, enhancing overall performance.
- An anomaly detection model is formulated for packet-level IDS to identify unknown novel attacks.
- To enhance the overall detection robustness, a sample agent with a complementary reward policy is introduced within the RL framework for adversarial training.
- Recognizing the sensitivity of RL to hyperparameter choices, this paper employs Bayesian search to determine optimal parameters, ensuring robust and effective intrusion detection.

The remainder of the paper is structured as follows: Sect. 2 provides a concise review of pertinent literature. Section 3 elaborate on the proposed intrusion detection frameworks. Section 4 presents the results obtained from experiments conducted at packet level. Lastly, Sect. 5 concludes the paper, offering insights and discussing potential avenues for future research.

2 Related Work

There has been growing interest in applying reinforcement learning to network security. Various approaches have been proposed, including Q-learning, deep reinforcement learning, and multi-agent reinforcement learning. Most of these approaches have focused on specific aspects of network security, such as intrusion detection, anomaly detection, or attack prediction.

One notable study is the work by Oussama et al. (2019), who applied Q-learning to develop an adaptive intrusion detection system. Their approach involved defining the state as a set of network features, such as the number of connections, the packet size, and the source IP address. The action was defined as whether to block or allow the incoming traffic, and the reward was defined as a function of the security level of the network. Their experiments showed that Q-learning can effectively learn an adaptive strategy for intrusion detection, and can outperform traditional rule-based approaches.

Previous research has delved into the utilization of machine learning, encompassing Deep Reinforcement Learning (DRL) [7,8], both supervised and unsupervised learning, has been conducted in previous work [9,10]. The authors of [9] presented a systematic overview of DRL regarding cybersecurity. They discussed the DRL-based security solutions for autonomous and non-autonomous physical systems as well as the game theory-based multi-agent DRL cyber-attack defense system.

The studies were conducted in light of real-world and virtual settings. In [10], the authors demonstrated how adversarial learning can help improve the robustness of the DRL-based IDS model used for cyber defense. In their approach, adversarial training was performed by randomly sampling a session from the dataset and generating a positive reward based on the prediction at first, and then the environment was forced to predict wrong. Doing this, the environment acts as an adversarial agent and causes the classifier to further learn better underlying features that ultimately improve the robustness of the DRL-based IDS system. In [7], the authors applied the NSL-KDD [11] and AWID [12] datasets to the study of many DRL algorithms for intrusion detection, including Double Deep Q-Network (DDQN), Deep Q-Network (DQN), Policy Gradient (PG), and Actor-Critical (AC) methods. The supervised machine learning algorithms were trained on those datasets to categorize intrusion occurrences. They used the idea of a virtual environment and sampling methods to replace the interaction module of RL training. It was demonstrated how DRL may outperform alternative machine learning methods when applied to today's data networks requiring immediate attention and reaction. The results show that DDQN is superior to the competing methods in terms of classification performance and learning speed.

Based on the stateful MDP, deep Q-learning was suggested in [13,14]. They contrasted their method's efficacy with conventional planning-based and DRL-based IDS, using metrics such as execution time, batch size, and cumulative reward to measure performance.

In [15], the authors advocated the deployment of a Reinforcement Learning (RL) agent on routers to learn from network traffic and redirect it away from the target server. They introduced the innovative concept of employing the Coordinated Team Learning (CTL) approach instead of the traditional multi-agent RL training. Their proposed algorithm demonstrated improved scalability in handling network traffic.

Additionally, Hidden Markov Models (HMMs) were introduced in [16] as a machine learning technique for detecting multi-step attacks by predicting the attacker's next move. The authors utilized preliminary offline training, where the Intrusion Detection System (IDS) prediction was compared with stored database information to prevent overfitting. In [17] a decision-theoretic framework named Anomaly Detection And Response System (ADRS) was presented, focusing on cost-sensitive and self-optimizing performance. The authors explored its functionality in autonomous networks. Furthermore, [18] showcased an effective IDS by merging multi-objective decision problems with evolutionary algorithms. They approached IDS as a multi-attribute decision-making challenge, considering factors like implementation cost, resource limitations, time efficiency, and modification costs. The objective was to minimize these functions before responding to threats. A preprocessing technique using a weighted linear combination method was employed to map diverse network traffic to a common metric, enhancing classification accuracy. In a recent study [19], the authors proposed an effective method for fine-tuning hyperparameters in DQ-learning. Their evaluation on the NSL-KDD dataset revealed that a lower exploration rate led to improved classification accuracy.

In the work presented by [20], an online version of Intrusion Detection System (IDS) using Deep Q-Networks (DQN) was introduced. The innovative approach involved incorporating auto-encoders into the Q-network of Reinforcement Learning (RL) as function approximators. This integration contributed to enhanced classification accuracy, particularly when processing real-world data through the RL-trained network. The utilization of auto-encoders significantly mitigated the impact of human interaction, leading to considerably higher efficiency during the online training mode.

Much of the prior research in Intrusion Detection Systems (IDS) has concentrated on enhancing performance or evaluating proposed strategies against similar Machine Learning (ML)-based approaches [21–23]. In alignment with the literature review, our objective in this paper is to introduce a comprehensive Reinforcement Learning (RL) framework for packet-level IDS challenges, emphasizing heightened robustness. The proposed method not only identifies malicious traffic but also categorizes it into distinct classes, including normal traffic and various types of attacks. To the best of the authors' knowledge, this holistic approach has not been extensively explored in existing literature, and its detailed exploration will be presented in the next section.

3 Methodology

3.1 Components

In this paper, we examine the application of Q-learning to formulate an optimal strategy for network defense. We define the state using network features, such as active connections, traffic type, and compromised devices. The action encompasses defensive measures like IP address blocking, traffic restrictions, and service shutdown. Additionally, the reward is determined by the network's security level after implementing the action. Our investigation aims to enhance network security by utilizing Q-learning to develop an optimal defense strategy against cyber attacks. We define the state, action, and reward, applying Q-learning to construct an Intrusion Detection System (IDS) agent for anomaly detection in network packets, as despicted in Fig. 1. Packets transmitted or received by network devices are intercepted, forwarded for data preprocessing and feature engineering, and subsequently stored in a database for training. The RL-based IDS system leverages this packet data to learn and identify anomalies. The acquired policy is then evaluated against real-world scenarios and deployed to automate decision-making or notify network administrators of anomaly occurrences.

The experiments demonstrate that Q-learning effectively learns an optimal defense strategy, surpassing traditional rule-based approaches in specific scenarios.

Our methodology revolves around three benchmark datasets: the KDD Cup 99 dataset, ISCX dataset, and CICIDS 2017 dataset, employed for constructing and comparing Network Intrusion Detection Systems (NIDS). We utilize classical supervised machine learning techniques, namely CNN, K-NN, and SVM, alongside the Double Deep Q-Network (DQN) with dueling architecture. These datasets have been widely adopted by researchers for simulating network IDS and have been referenced in numerous studies over the past two decades.

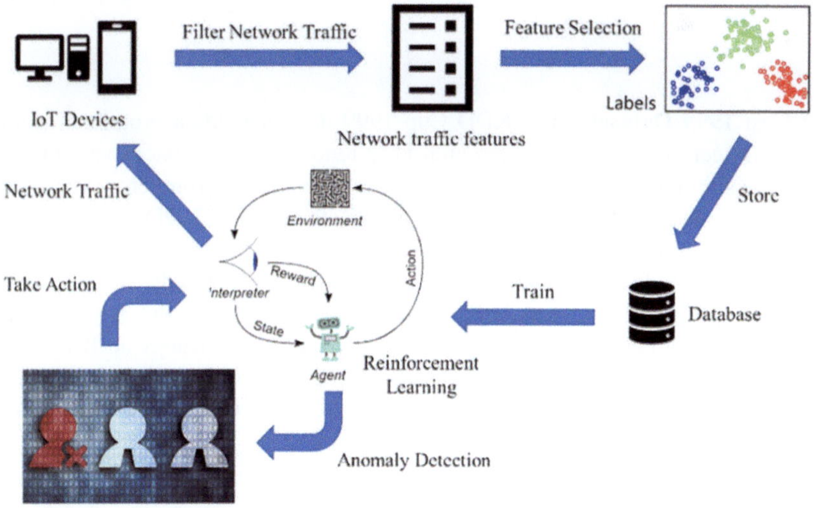

Fig. 1. DQN with dueling-based IDS.

The ISCX dataset comprises synthetically recorded packets spanning seven days, replicating real-time network traffic. On the other hand, the KDD Cup 99 dataset encompasses seven weeks of network traffic derived from raw TCP dump data, mimicking a typical United States Air Force local-area network (LAN). The CICIDS 2017 dataset, a more recent addition, provides real-world network traffic scenarios with modern attack patterns, offering a more comprehensive evaluation of IDS models. Our methodology incorporates a data analytics approach, encompassing data exploration and preprocessing, as well as the training and evaluation of machine learning models.

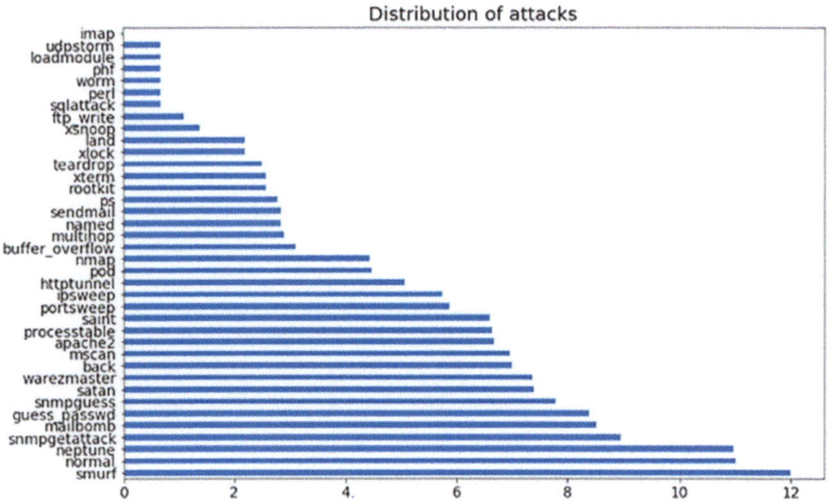

Fig. 2. KDD Cup 99 Dataset.

KDD Cup 1999 Dataset: The KDD Cup 1999 dataset holds a prominent status in Intrusion Detection System (IDS) research [11], renowned for its extensive data volume and diverse simulated intrusions within a military network environment.

Described in detail by Lopez-Martin et al. [5], the original KDD Cup 1999 dataset comprises 42 features, with 34 being continuous and the remainder represented as symbolic features. The dataset encompasses a total of 4,898,431 records, covering 7 weeks of network traffic. For our research, we extract samples from the original dataset, creating a training dataset of 311,029 samples and a testing dataset of 494,021 samples, with no overlapping samples between them. Unlike previous binary classification approaches, our implementation focuses on multi-class classification, considering all labels during the training process.

In Fig. 2, the frequency of classes in the KDD Cup 1999 is depicted, with the number of records transformed to a logarithmic scale for enhanced visualization of the imbalanced training and testing datasets. Table 1 provides a summary of attacks categorized into four groups for both the training and testing datasets.

ISCX 2012 Dataset: The ISCX 2012 dataset is composed of labeled network traces, encompassing full packet payloads in the pcap format (Shiravi et al., 2012). Captured over a week in an emulated network environment, the dataset includes both normal and malicious network behaviors. It is structured around profiles, containing intrusion descriptions and abstract distribution models for applications, protocols, or network entities. The ISCX 2012 dataset comprises 481,677 records for training and 160,519 records for validation and testing. Figure 3 illustrates the class frequencies in the ISCX 2012 training and test datasets.

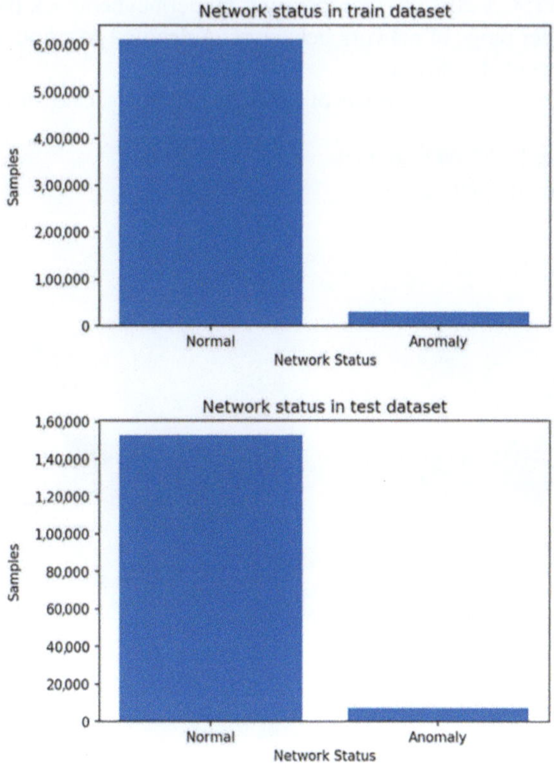

Fig. 3. ISCX 2012 dataset.

Table 1. IDS Framework at Packet level

Module	Sub-Module	Function
Preprocessing	None	Data Transformation and Feature Engineering
Reinforcement Learning	Interactive Training	Store batch data into replay buffer
Anomaly Detection	None	Detect attacks blind to the training module

CICIDS 2017 Dataset: The Canadian Institute for Cybersecurity Intrusion Detection System (CICIDS) 2017 dataset is one of the most widely used datasets for network intrusion detection system (NIDS) research. It was created to simulate real-world network traffic, containing both benign and malicious activities [24].

This dataset was designed to overcome the limitations of older datasets such as KDD Cup 99 and ISCX 2012 by incorporating modern cyberattack patterns, real-time traffic, and a broader range of network behaviors. A detailed description of the CICIDS 2017 dataset is provided in Fig. 4.

The CICIDS 2017 dataset consists of 2,830,743 samples, categorized as follows:

- **Benign Traffic**: 2,273,097 samples
- **Attack Traffic**: 557,646 samples

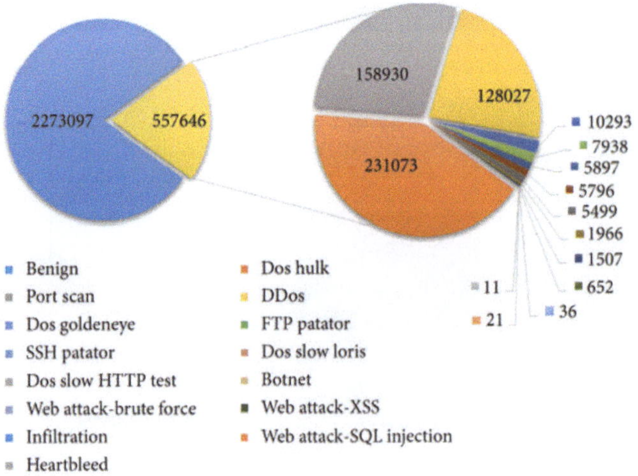

Fig. 4. CICIDS 2017 dataset.

3.2 The Proposed IDS Framework at the Packet Level

In the realm of packet-level intrusion detection, the extraction of messages and headers from network packets takes precedence for identifying malicious traffic. A packet's structure includes messages from the application layer and three headers from the network layers. The content of application layer messages varies with different protocols; for instance, HTTP generates specific HTTP request messages. Additional information is carried by various headers, such as TCP and UDP headers for the prevalent transportation layer protocols, IP headers for the network layer, and Ethernet headers for the Physical/Ethernet layer. The knowledge encapsulated within these headers is referred to as a field.

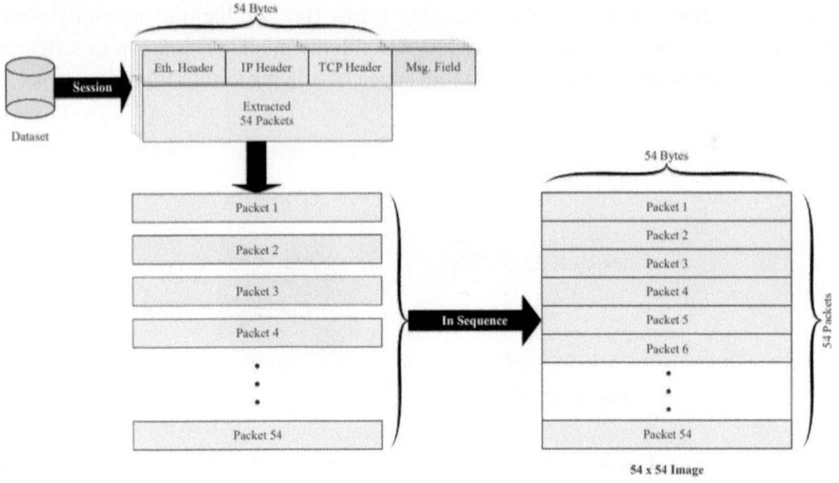

Fig. 5. Procedure of the proposed image embedding [25].

The proposed Intrusion Detection System (IDS) framework at the packet level is delineated in Table 1, featuring two major modules and several sub-modules. The Preprocessing module is designed for data transformation and feature engineering, while the Reinforcement Learning module is dedicated to training the IDS using Reinforcement Learning (RL) approaches. This module comprises a training module and an interaction module. Furthermore, an additional anomaly detection module is introduced in this paper to identify novel attacks not covered by the training module.

- Preprocessing Module: Taking inspiration from the application of natural language processing (NLP) techniques like word2vec for handling extensive text documents and metadata files, this section introduces an embedding method called image embedding for the RL-based Intrusion Detection System (IDS). Procedure of the proposed image embedding, transforming packets to an image are despicted in Fig. 4 and 5. Table 2 outlines the key steps of the preprocessing module.

- Reinforcement Learning Module: Before developing a reinforcement l module, it is necessary to transfer the intrusion detection problem into an RL-based problem. Opting for DQN learning proves to be an excellent solution for addressing discrete problems, and its implementation is considerably more straightforward compared to Proximal Policy Optimization (PPO). With these crucial elements of reinforcement learning identified, we proceed to craft our reinforcement learning module. Convolutional Neural Networks (CNNs) are selected as the network structure for the IDS agent, given that the input states are in the form of images. It's worth noting that 1D-CNN is also suitable for the IDS. The images for each session are treated as a form of time-series data, as the packets embedded in the image are organized chronologically based on their capture time.
- Anomaly Detection Module: The anomaly detection module aims to identify novel attacks that elude detection in the training set by treating them as anomaly classes. This is crucial for a robust Intrusion Detection System (IDS) since, in practical scenarios, it is impractical to encompass all types of attacks within the training set.

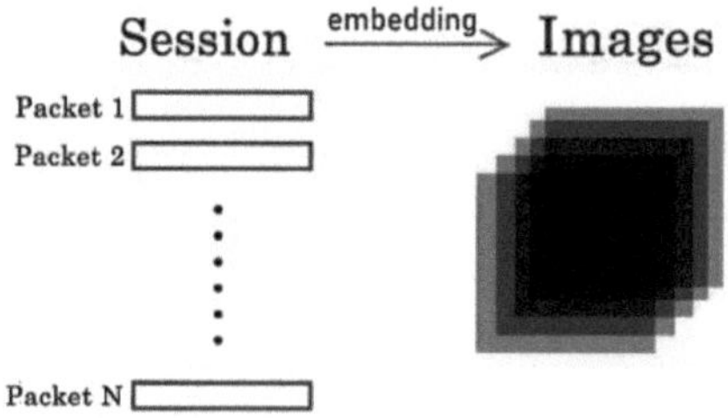

Fig. 6. Transform a session into a batch of images.

Table 2. Preprocessing Module at Packet level

Step	Input	Method	Output
1	Raw Network Traffic	Session-based Rule	Seperated Sessions
2	Seperated Sessions	Image Embedding	Session Images
3	Session Images	Labeling with Log Files	Images and Label
4	Images and Label	Normalization	Applicable Dataset

3.3 Reinforcement Learning Module

Prior to constructing a reinforcement learning module, it is essential to transform the intrusion detection problem into an RL-based problem.

Algorithm 1 Interaction Module

procedure START INTERACTING
top:
  Randomly sample a session and obtain its label
  Take the first image in the session as current state s_1
  Feed s_1 into agent and obtain action/prediction a_1
  Feed a_1 and label into reward mechanism, obtain reward r_1
  $i \leftarrow 0$
loop:
  $i \leftarrow i+1$
  if Last image in the session **then**
    Store $(s_i, a_i, r_i, None)$ into replay buffer
    break
  else
    Take next image in the session as next state s_{i+1}
    Store (s_i, a_i, r_i, s_{i+1}) into replay buffer
  end if
  goto *loop*
  goto *top*
end procedure

Fig. 7. Reinforcement Learning Module.

The RL agents take a finite batch of images as input states. CNN and 1D-CNN are utilized for feature extraction in DQN-CNN and DQN-1D-CNN, respectively. An extra fully connected layer is employed for the ultimate classification and detection. The output consists of Q-values for the current state, and the decision for prediction/action. The agent can follow the trajectory until it reaches the end (last packet) of a session: s_0, a_1, r_1, s_1, a_2, r_2, ..., s_i, a_i, r_i, ..., s_t., as shown at Fig. 6.

Table 3. Training and Validation Set Details

Traffic	Training Images	Validation Images	Class
Normal	25,012	6256	0
PortMap	20,200	5058	1
NetBIOS	23,652	5912	2
LDAP	18,019	5083	3
MSSQL	22,986	4709	4
UDP	19,159	6051	5
UDP-Lag	19,552	5198	5
SYN	22,640	4970	7

The comprehensive process of the training module is illustrated in Fig. 7, using DQN as an example. Two networks, namely the Update and the Target, collaborate to accomplish the approximate regression. The configuration of the update network mirrors that of the agent, while the target network replicates the structure of the update network and is initialized with identical parameters. For instance, once the replay buffer is full, training begins for a specified number of epochs. Following the training phase, the new agent engages with the environment, storing fresh data into the replay buffer while removing the outdated data. The next section will present the experimental outcomes of Q-Learning for IDS, utilizing the two datasets mentioned above and incorporating enhancements to the algorithm.

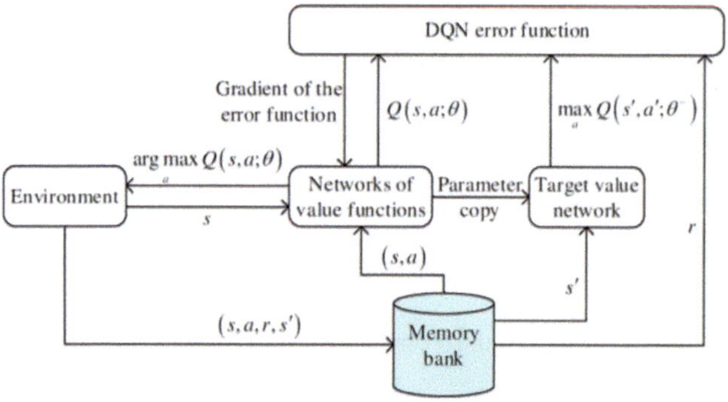

Fig. 8. Deep Q-Learning Training Module.

3.4 Anomaly Detection Module

The anomaly detection mechanism becomes particularly valuable when faced with novel attacks that were not present in the training dataset. Traditional supervised learning approaches may struggle to handle these unforeseen threats effectively. However, the Anomaly Detection Module, by treating instances with confidence scores (Q-values) below the specified threshold as anomalies, enables the system to recognize and respond to previously unseen attack patterns.

This adaptive capability is crucial for Intrusion Detection Systems (IDS) powered by Q-learning, as it acknowledges the dynamic nature of cyber threats. In practical scenarios, it is challenging to anticipate and include all potential attack types during the training phase. The Anomaly Detection Module addresses this limitation by allowing the system to flag deviations from normal behavior, contributing to a more resilient and effective defense against a broad spectrum of cyber threats.

Furthermore, the determination of anomalies based on the confidence scores provides a mechanism for real-time decision-making. When the Q-value falls below the threshold, the input session image is labeled as an 'anomaly' attack, triggering appropriate response measures. On the other hand, the class associated with the maximum confidence score offers insights into the expected detection result, providing valuable information for understanding and responding to potential threats.

In summary, the Anomaly Detection Module in Q-learning significantly enhances the adaptability and effectiveness of the Intrusion Detection System by enabling it to identify and respond to novel attacks that may not have been encountered during the training phase. This dynamic approach contributes to a more robust defense against evolving cyber threats in real-world environments.

3.5 Experimental Setup

Our methodology revolves around three benchmark datasets: the KDD Cup 99 dataset, ISCX dataset, and CICIDS 2017 dataset, employed for constructing and comparing Network Intrusion Detection Systems (NIDS). We utilize classical supervised machine learning techniques, namely CNN, K-NN, and SVM, alongside the Double Deep Q-Network (DQN) with dueling architecture. These datasets have been widely adopted by researchers for simulating network IDS and have been referenced in numerous studies over the past two decades.

The ISCX dataset comprises synthetically recorded packets spanning seven days, replicating real-time network traffic. On the other hand, the KDD Cup 99 dataset encompasses seven weeks of network traffic derived from raw TCP dump data, mimicking a typical United States Air Force local-area network (LAN). The CICIDS 2017 dataset, a more recent addition, provides real-world network traffic scenarios with modern attack patterns, offering a more comprehensive evaluation of IDS models. It consists of a mixture of normal and malicious traffic captured in a realistic network environment, including attacks such as brute force, botnets, denial-of-service (DoS), infiltration, and web-based attacks. The dataset is structured to reflect contemporary cyber threats, making it highly relevant for evaluating modern IDS solutions. Our methodology incorporates a data analytics approach, encompassing data exploration and preprocessing, as well as the training and evaluation of machine learning models.

3.6 Computational Cost Analysis

One major concern in deploying RL-based IDS is the computational cost. We analyze:

- **Training time**: Measured in epochs across different datasets.
- **Memory consumption**: The required RAM and GPU usage during model training.
- **Scalability**: The performance of the IDS when applied to large-scale network environments.

4 Experimental Results

Our experiments demonstrate the effectiveness of Q-learning in acquiring an optimal strategy for network defense. We conduct a performance comparison between the Q-learning algorithm and traditional rule-based approaches like firewalls and intrusion detection systems. The training set encompasses eight distinct traffic types, namely Normal, PortMap, NetBIOS, LDAP, MSSQL, UDP, UDP-Lag, and SYN, despicted in Fig. 8. The test set comprises 12 traffic types. In addition to the eight types present in the training set, four additional types – NTP, DNS, WebDDoS, and TFTP – are included in the test set. The inclusion of these extra traffic types contributes to the IDS's generality as they were not part of the training set. In the proposed approach, an anomaly detection model identifies these four additional attacks by treating them as an anomaly class.

The results indicate that Q-learning can surpass traditional approaches in specific scenarios, especially in dynamic and complex network environments. To assess the proposed packet-level IDS, we train two agents based on DQN-CNN and DQN-1D-CNN. Additionally, two agents, PG-CNN and PG-1D-CNN, are trained using the policy gradient learning method for comparison. The dataset is divided into a 4:1 ratio for the training and validation sets. The session length distribution remains consistent between the training and validation sets, ensuring an approximate image ratio of 4:1. Following the partition, the dataset statistics are presented in Table 3.

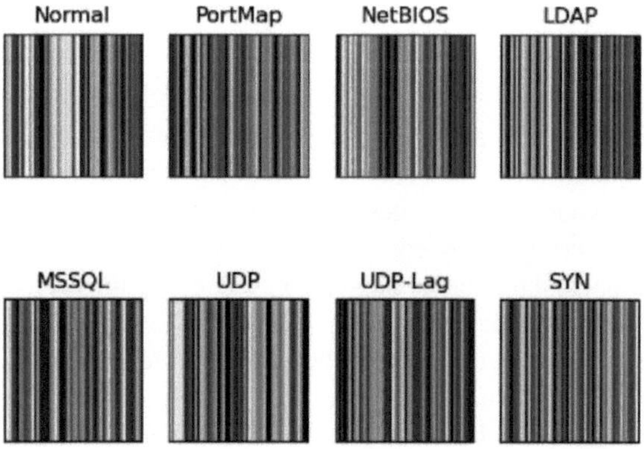

Fig. 9. Images after image embedding for each class.

In the experiment, we compare the performance of four agents using three reward discount values: 0.1, 0.5, and 0.9. The evaluation is based on the accuracy averaged over each class on the validation set to determine the optimal discount value. The results are presented in Table 4. Notably, when reward discount is set to 0.1, all four agents exhibit the highest performance. In terms of reinforcement learning algorithms, DQN-1D-CNN slightly outperforms PG-1D-CNN across all discount factors, and DQN-CNN also marginally outperforms PG-CNN across the board. Furthermore, it is observed that PG-based agents demonstrate higher sensitivity to discount factors. The overall results suggest that Deep Q-Learning is a comparatively superior algorithm to policy gradient methods in our experiments. Therefore, in the subsequent case studies, we assume a fixed discount factor value of 0.1 and primarily employ Deep Q-Learning algorithms for the agent.

Detailed experiment results for intrusion detection using DQN-1D-CNN with a discount value of 0.1 are presented in Table 5. As indicated in the table, our detection system achieves a high accuracy of 98.78%. In terms of individual classes, 100% of normal traffic is successfully detected. The Recall (detection rate) for MSSQL is the

Table 4. Performances on four agents

Agent	Discount Factor	Validation Accuracy
DQN-1DCNN	0.1	98.78%
	0.5	95.21%
	0.9	87.12%
DQN-CNN	0.1	96.07%
	0.5	93.22%
	0.9	86.24%
PG-1DCNN	0.1	96.17%
	0.5	90.15%
	0.9	79.67%
PG-CNN	0.1	96.03%
	0.5	87.12%
	0.9	70.69%

Table 5. Results of DQN-1DCNN and DQN-CNN

Agent	Traffic	Accuracy	Precision	Recall	F1-Score
DQN-1DCNN	Normal		99.32%	100%	99.66%
	PortMap		98.04%	98.95%	98.49%
	NetBIOS		98.87%	99.26%	99.06%
	LDAO	98.78%	98.85%	98.37%	98.61%
	MSSQL		98.47%	97.30%	97.86%
	UDP		98.95%	99.62%	99.26%
	UDP-Lag		98.02%	98.31%	98.61%
	SYN		98.64%	97.83%	98.23%
DQN-CNN	Normal		96.91%	99.68%	98.27%
	PortMap		95.67%	93.53%	94.59%
	NetBIOS		97.35%	95.75%	96.55%
	LDAO	96.07%	95.76%	96.95%	96.35%
	MSSQL		95.06%	96.03%	95.54%
	UDP		95.32%	96.00%	95.66%
	UDP-Lag		95.05%	94.92%	94.99%
	SYN		97.14%	94.87%	95.71%

lowest among the eight types but still impressive at 97.30%. Table 5 also provides the validation results for DQN-CNN, revealing slightly lower performance compared to DQN-1D-CNN but still achieving a commendable accuracy of 96.07%. These experimental findings demonstrate that both DQN-1D-CNN and DQN-CNN exhibit high performance in handling image-related tasks. Furthermore, in our experiments, treating the image as time series data is effective, and in this scenario, DQN-1D-CNN demonstrates superior performance. Additionally, the average computation times for intrusion detection using DQN-CNN and DQN-1D-CNN are provided. In the context of packet-level approaches, a comparison is made between the proposed method and conventional 1D-CNN and CNN approaches (without RL). It is evident that conventional deep learning methods can attain relatively high performance on the test set, indicating that image embedding serves as a valuable data preprocessing technique for raw network traffic analysis. However, the crucial observation is that the proposed RL-based approach exhibits significantly superior generalizability when compared to these traditional deep learning approaches.

5 Conclusions

This paper introduces a packet-level Intrusion Detection System (IDS) utilizing Reinforcement Learning (RL). Employing image embedding, network traffic undergoes transformation into images processed by both CNN and 1D-CNN for packet-level detection. Furthermore, an anomaly detection module enables the identification of unknown network traffic. Experimental results confirm the superior performance of the proposed packet-level compared to commonly used state-of-the-art algorithms. However, further research can be pursued based on the following future directions.

- The concept of exploration can be incorporated into the packet-level reinforcement learning framework, utilizing an exploration policy to encompass a broader range of states in the interaction space.
- While the proposed packet-level IDS can successfully detect malicious images, it faces challenges in pinpointing the specific position of the traffic session embedded in the image. One potential solution is to employ an N-to-N LSTM model to address this particular issue.
- Finally, for additional validation of the proposed IDS design, experiments can be executed on diverse datasets featuring various types of attacks. A comprehensive comparison can then be carried out with other designs of RL-based IDS.

In intricate and unpredictable environments, constructing an Intrusion Detection System (IDS) with Deep Reinforcement Learning (DRL) presents advantages in terms of computational efficiency, requiring minimal updates compared to traditional machine learning algorithms. Additionally, DRL offers an advantage in hyper-parameter tuning, as minimal updates are needed to enhance performance.

Subsequent research endeavors could focus on creating a multi-class classification model to discern the specific attack type upon detecting malicious activity. Moreover, exploring contemporary datasets for IDS development could provide insights into recently evolved attack types not covered by the existing datasets, namely KDD Cup 99 and ISCX 2012, published in 2009 and 2012, respectively.

References

1. Hu, Z., Chen, P., Zhu, M., Liu, P.: Reinforcement learning for adaptive cyber defense against zero-day attacks, 2019
2. Larimer, D.: Dpos consensus algorithm - the missing white paper. https://steemit.com/dpos/@dantheman/dpos-consensus-algorithm-this-missing-white-paper, 2018, 2017
3. Badr, Y.: Enabling intrusion detection systems with dueling double deep q -learning. Digital Transformation and Society, 1, 2022
4. Khoei, T.T., Kaabouch, N.: Acapsule q-learning based reinforcement model for intrusion detection system on smart grid. volume , 2023
5. Lopez-Martin, M., Carro, B., Sanchez-Esguevillas, A.: Application of deep reinforcement learning to intrusion detection for supervised problems. Expert Systems with Applications, 141, 2020
6. Liu, Z., Yin, X., Hu, Y.: Cpss lr-ddos detection and defense in edge computing utilizing DCNN q-learning. IEEE Access, 8, 2020
7. François-Lavet, V., Henderson, P., Islam, R., Bellemare, M.G., Pineau, J.: An introduction to deep reinforcement learning. Foundations and Trends in Machine Learning, 11, 2018
8. Sethi, K., Madhav, Y.V., Kumar, R., Bera, P.: Attention based multi-agent intrusion detection systems using reinforcement learning. J. Inf. Secur. Appl. 61, 2021
9. Nguyen, T.T., Reddi, V.J.: Deep reinforcement learning for cyber security. IEEE Trans. Neural Netw. Learn. Syst. 2021
10. Caminero, G., Lopez-Martin, M., Carro, B.: Adversarial environment reinforcement learning algorithm for intrusion detection. Comput. Netw. 159, 2019
11. Tavallaee, M., Bagheri, E., Lu, W., Ghorbani, A.A.: A detailed analysis of the KDD cup 99 data set. 2009
12. Kolias, C., Kambourakis, G., Stavrou, A., Gritzalis, S.: Intrusion detection in 802.11 networks: Empirical evaluation of threats and a public dataset. IEEE Communications Surveys and Tutorials, 18, 2016
13. Iannucci, S., Barba,O.D., Cardellini, V., Banicescu, I.: A performance evaluation of deep reinforcement learning for model-based intrusion response. 2019
14. Iannucci, S., Cardellini, V., Barba,O.D., Banicescu, I.: A hybrid model-free approach for the near-optimal intrusion response control of non-stationary systems. Future Generation Computer Systems, 109, 2020
15. Malialis, K., Kudenko, D.: Distributed response to network intrusions using multiagent reinforcement learning. Eng. Appl. Artif. Intell. 41, 2015
16. Holgado, P., Villagra, V.A., Vazquez, L.: Real-time multistep attack prediction based on hidden markov models. IEEE Trans. Dependable Secure Comput. 17, 2020
17. Zhang, Z., Naït-Abdesselam, F., Ho, P.H., Kadobayashi, Y.: Toward cost-sensitive self-optimizing anomaly detection and response in autonomic networks. Computers and Security, 30, 2011
18. Fessi, B.A., Benabdallah, S., Boudriga, N., Hamdi, M.: A multi-attribute decision model for intrusion response system. Inf. Sci. 270, 2014
19. Alavizadeh, H., Alavizadeh, H., Jang-Jaccard, J.: Deep q-learning based reinforcement learning approach for network intrusion detection. Computers, 11, 2022
20. Kim, C., Park, J.S.: Designing online network intrusion detection using deep auto-encoder q-learning. Comput. Electr. Eng. 79, 2019
21. Aljehane, N.O., et al.: Golden jackal optimization algorithm with deep learning assisted intrusion detection system for network security. Alexandria Eng. J. 86, 2024
22. Alqahtani, J., Akram, M., Ali, G.A., Iqbal, N., Alqahtani, A., Alroobaea, A.: Elevating network security: A novel s-box algorithm for robust data encryption. IEEE Access, 12, 2024

23. Alwis, C.D., Porambage, P., Dev, K., Gadekallu, T.R., Liyanage, M.: A survey on network slicing security: attacks, challenges, solutions and research directions. IEEE Communications Surveys and Tutorials, 26, 2024
24. Kurniabudi, Stiawan, D., Darmawijoyo, Bin Bin Idris, M.Y., Bamhdi, A.M., Budiarto, R.: Cicids-2017 dataset feature analysis with information gain for anomaly detection. IEEE Access, 8, 2020
25. Thanh, C.T.: A novel approach for intrusion detection based on deep belief network. volume 1225 AISC, 2020

Web Network and Mobile App Authentications Enhancing Social Media Security

Rachel John Robinson[1(✉)] and Aaron Taki[2]

[1] IU Hochshule, Frankfurter Alle 73A, 10247 Berlin, Germany
info@rachel-johnrobinson.com
[2] Arden University, Berlin Campus, 10963 Berlin, Germany
ataki@arden.ac.uk

Abstract. The proliferation of the internet has played a pivotal role in altering the way people communicate and interact in today's world. Significantly, social media platforms have become a crucial component of the ever-changing digital environment, influencing the sharing of information, social connections, and commercial transactions. Due to this, social media has expanded beyond its conventional function of connecting individuals to generating economic prospects for individuals and organizations. Moreover, the uninterrupted connection enabled by these platforms has fundamentally transformed social interactions, completely overhauled marketing tactics, and additionally established enhanced opportunities for interaction. A study carried out in 2020 found that over 50% of the global population actively engages with social media platforms. This trend is expected to continue growing as internet services become more accessible and affordable. Regrettably, the current prevalence of users on social media platforms has both advantages and disadvantages. These challenges encompassed security and data privacy concerns arising from authentication system issues. Therefore, the current research seeks to examine the intricate domain of authenticating social media online and mobile applications. To accomplish this, the study will specifically address the difficulties associated with preserving user privacy and convenience, while also ensuring secure access to user accounts. This study aims to further the development of social media authentication systems by analysing current procedures, detecting vulnerabilities, and assessing technology.

Keywords: social media · privacy · vulnerability · authentication

1 Introduction

1.1 Proliferation of the Social Media

Social media platforms have become essential tools for facilitating relationships between individuals, spreading information, and fostering worldwide communities. Scholars have analysed the significant influence of social media on the evolution of modern communication patterns, considering the active involvement of billions of users on platforms such as Facebook, Instagram, LinkedIn, and Twitter. It is [1] provided a definition of

L. Meng (Ed.): CCCN 2025, SCT, pp. 127–141, 2025.
https://doi.org/10.1007/978-3-032-03632-2_10

social media (also known as social networking sites) as a web-based service that facilitates communication between persons via the internet. Moreover, to provide a more distinct illustration, social media can be classified into various groups according to their primary function in serving the community. Platforms such as Wikipedia, which aim to supply diverse information, can be classified as collaborative projects while platforms like X can be characterized as blogs or microblogs. Conversely, Facebook, LinkedIn, and MySpace are classified as social networking platforms, whereas YouTube and Flickr constitute content communities [2].

In terms of global usage, these platforms command different numbers based on their classification. For instance, being a social networking platform, Facebook has close to 3 billion monthly active users [3]. Content platforms such as YouTube, Instagram, and TikTok have 2.5 billion, 2 billion, and 1.05 billion active monthly users respectively [3]. Collectively, these platforms engage close to 60% of the global population, with a majority being 18–34 years. The definitions and categorizations of social media platforms are crucial for understanding the extensive presence of this technology in contemporary culture.

1.2 Common Social Media Authentication Processes

Authentication refers to the process of verifying the identity of individuals who are attempting to access digital platforms. It is a crucial measure for safeguarding user accounts and avoiding unauthorized access and malicious misuse of sensitive data [4]. As social media platforms continue to evolve and expand, there is an increasing need for robust authentication techniques that are specifically tailored to their unique requirements. As a result, the prominent social media platforms implemented and embraced enhanced and user-centric authentication techniques, including two-factor authentication (2FA), biometric authentication, behavioral biometrics, Single Sign-On (SSO), and security keys.

To state that the two-factor authentication approach necessitates users to present two separate authentication factors for their identities to be validated. By employing this approach, creators of social media platforms and mobile applications enhance security measures, making it more challenging for attackers to infiltrate a user's device or online account. Regarding market utilization, it is emphasized that the use of two-factor authentication (2FA) witnessed a growth of over 50% from 2017 to 2021 [5]. Although the strategy effectively prevents automated bots, phishing attempts, and targeted attacks, it also raises the risk of account lockout if the user does not have access to a phone or second email address. Figure 1 below depicts the concept of 2-factor Authentication.

Another often employed kind of authentication is biometric, as described [6]. This technique utilizes biological features such as fingerprints, voices, retinas, and facial features to grant access. Techniques such as fingerprint scanners, facial recognition, and iris scanners can be employed to assess the distinctive characteristics required to verify the individual. Due of the distinctiveness of these qualities for each person, it is highly challenging for attackers to imitate them and circumvent the authentication mechanism. According to [6], the presence of a crucial characteristic has led to the widespread adoption of biometric applications, resulting in millions of downloads. Notably, firms like Apple have incorporated these applications into their access and payment systems.

Fig. 1. Two Factor Authentication

Behavioral authentication is a crucial method for identifying users by analysing their interactions with their gadgets, such as their walking style, mouse movements, and typing patterns [7]. SSO, or Single Sign-On, enables users to authenticate once and obtain access to all systems without requiring them to log in separately for each system [7].

Given the significant weaknesses and risks associated with traditional methodologies, stakeholders have shown a strong interest in using emerging technologies to establish secure and reliable means of authentication. It is to emphasize that blockchain is renowned for its decentralized and transparent characteristics and is now being employed to enhance systems beyond cryptocurrency. IT professionals are actively searching for ways to leverage the capabilities of blockchain technology to tackle cybersecurity issues. To state that blockchain-based authentication enhances security for consumers through the utilization of cryptographic methods and decentralized networks [8].

1.3 Specific Objectives

The main aim of this study is to improve the security of social media websites and native mobile applications and lessen frequent security risks this research attempts to investigate and suggest sophisticated authentication mechanisms.

To achieve this objective, the study will employ the following specific objectives.

1. To examine the current methods of authentication used in popular social media platforms.
2. To identify specific security risks and vulnerabilities associated with these existing authentication mechanisms in social media and mobile apps.
3. To investigate cutting-edge methods of authentication like blockchain-based authentication and their role in enhancing security within digital platforms.

To evaluate the efficacy of the proposed Blockchain-based model in improving trust, privacy, and security on social media platforms.

Ultimately, the researcher aims to advance the field of social media web and mobile application authentication by addressing new opportunities and security and usability concerns in the context of quickly changing technical environments. The research provides novel insights and solutions to improve authentication processes in social media contexts by examining state-of-the-art authentication technology, emphasizing user-centric design principles, and looking into cross-platform authentication difficulties. This will not only promote technological innovation but also contribute to the academic knowledge.

2 Literature Review

As such, this chapter aims to explore the current body of knowledge surrounding the authentication methods in the past, and present and how they relate within the social media landscape. Using peer-reviewed journals from Google Scholar, research databases and reports from accredited institutions such as the General Data Protection Regulation (GDPR), the researcher synthesizes the different input provided by the scholars.

2.1 Technology Acceptance Model (TAM)

The TAM can be used to explain the surge in the use of social media platforms among the modern population and also how it relates to the authentication security processes. To begin with, a study by [9] showed that there was a clear relationship between user perceptions and frequency of use related to social media networks such as Facebook, X, and Pinterest. The study further established that there was a positive relationship between PU and the frequency of social media use, suggesting that when users view them perceive a social media platform to be useful, they are likely to spend more time using it. Similarly, a study by [9] that focused on Facebook, Myspace and X, established that PU and PEOU had much significance on the intention of use in social media. Similarly, [10] noted that TAM has been applied to various social media platforms. For instance, they explored user acceptance of Facebook features, privacy settings, and authentication methods. Besides, based on these feedback platforms such as Twitter, LinkeIn and Instagram have constantly evolved to improve security practices and align with the changing user behaviour.

2.2 Communication Privacy Management Theory (CPM)

In the context of authentication security, the CPM offers insights into users' privacy concerns and disclosure behaviours within digital platforms. Precisely, the theory sheds light on how users engage in privacy management strategies to control access to their personal information and protect their privacy rights. This may also entail the decision regarding the disclosure of sensitive information during the authentication process, such as passwords, biometric data and personal identifiers. Similarly, CPM can also provide insight into how employing sophisticated approaches such as blockchain-based authentication can switch social media users' privacy management practices [11].

2.3 Multifactor Authentication (MFA)

To address the shortcomings of password-based authentication, multifactor authentication (MFA) systems were introduced. According to [11], the MFA combines two or more authentication factors to verify a user's identity, thereby enhancing security. Similar studies by [12] indicate that MFA was a key element of a strong identity and access management policy as it created layers that needed to be verified thereby decreasing the tendency of a successful cyber-attack. The MFA authentication methodology assumes three types of additional information including the password or PIN which is known to the user, something the user has like a token or smartphone and something the user has like the fingerprint or facial recognition. It is noted that by requiring different authentications, the MFA systems reduce the likelihood of unauthorized access, even if one factor is compromised. Figure 2 below illustrates the MFA.

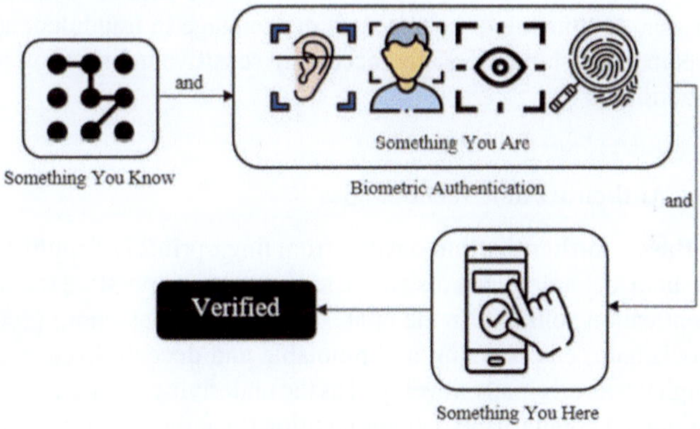

Fig. 2. Multi-Factor Authentication

While the MFA significantly enhances the security of social media platforms and other media platforms, it has some limitations. For instance, [13] note that MFA remains vulnerable and susceptible to phishing attacks, where cybercriminals trick users into revealing their credentials. Besides, also to indicate that the majority of the successful attacks on the MFA were orchestrated through sim-jacking, MFA hammering and AiTM attacks which compromise the effectiveness of the authentication process.

2.4 Security Risks and Vulnerabilities

Password Theft. According to [14], password theft remains to be a significant threat to the social media authentication process. Through this approach, attackers employ different techniques such as phishing attacks, malware and data breaches to steal user passwords. In light of these, the passwords extracted by the cyber attackers can then be used to get into the social media platform thereby accessing the sensitive information. For instance, it is noted that in 2012, LinkedIn suffered a massive data breach where hackers stole hashed passwords, leading to millions of accounts being compromised.

Credential Stuffing Attacks. Credential stuffing attacks involve the reuse of passwords across multiple accounts by attempting to log into numerous websites using stolen credentials obtained through data breaches. According to [15] during the credential stuffing attacks, the attackers automate the login process using botnets to rapidly test stolen username-password combinations until they find a match. In a similar instance, automated attacks were made on X in 2018, where the attackers used software to guess credentials and gain access to multiple accounts [15].

Third-party integrations and API Vulnerabilities. In the era of performance and AI, many digital platforms depend on third parties to enhance their functionality and user experience. However, [16] note that vulnerabilities in these integrations can expose user data to unauthorized access or manipulation by attackers.

Account Takeover Exploits. On the other hand, account takeover exploits involve using stolen credentials to gain access to a user account. To note that the attackers can access the login credentials, log into the account and change the account settings to lock out the legitimate user. At this point, the attackers may engage in fraudulent activities such as spreading spam or phishing messages, accessing sensitive information or conducting financial transactions [17].

2.5 Emerging Authentication Technologies

Block-Chain Based Authentication- Apart from fingerprints and multi-factor authentication, blockchain technology has also attracted interest as a possible social networking platform authentication solution. In the context of Bitcoin, Nakamoto (2008) proposed the idea of blockchain, emphasizing its immutable and decentralized structure. However, even though it was originally developed as the underlying infrastructure for Bitcoin, blockchain has gained popularity as a viable solution for authentication security. According to [18], blockchain provides a decentralized and tamper-resistant ledger that securely stores authentication credentials in a distributed network of nodes.

Decentralised Identity Management Systems- Further noted that decentralized identity management systems offer an alternative approach to traditional identity verification methods that rely on centralized authorities and proprietary databases. These systems enable users to control their digital identities and authenticate themselves without depending on centralized intermediaries. Thus, leveraging distributed ledger technology and cryptographic techniques, decentralized identity management systems provide a secure and privacy preserving framework for managing identity information [18].

2.6 Literature Gap

The identified gap in the above literature work is Lack of comprehensive analysis of emerging authentication technologies.

While the section mentions various emerging authentication technologies, such as blockchain-based authentication, decentralized identity management systems, and AI-powered authentication methods, it does not provide a comprehensive analysis of these technologies. The researcher could have explored more in-depth discussions on the benefits, limitations, and potential applications of each technology.

Some specific areas where further research could be conducted include:

- A comparative analysis of different blockchain-based authentication protocols to evaluate their security features and usability

- An examination of decentralized identity management systems' scalability, usability, and user experience
- An investigation into the use cases and implementation strategies for AI-powered authentication methods in social media platforms
- A review of existing research on the security and reliability of emerging authentication technologies

By addressing this literature gap, the researcher would provide a more comprehensive understanding of the current state of social media authentication and identify future directions for innovation and improvement.

3 Methodology

The main aim of this study was to comprehensively examine the authentication security within popular social media platforms and mobile applications. To accomplish this, the study specifically addressed the difficulties associated with preserving user privacy and convenience, while also ensuring secure access to user accounts. This study further aimed to explore the development of social media authentication systems by analysing current procedures, detecting vulnerabilities, and assessing technology. As such, this chapter discusses the research methodology employed to enhance the security of social media platforms by investigating current authentication methods, identifying associated security risks, exploring blockchain-based authentication, and evaluating its trust, privacy, and security implications.

3.1 Data Selection and Collection

This study aims to comprehensively examine the authentication security within popular social media platforms and mobile applications.

Methodology is to achieve this objective, the research will collect data from three primary groups: social media users, developers, and IT professionals. The study targets the following sub-groups:

- Social media users
- Developers
- IT professionals

To ensure a diverse range of perspectives, both random and purposive sampling methods will be employed.

Sampling Strategy: Purposive sampling will be used to select cybersecurity experts and IT professionals who are knowledgeable about the research topic. These individuals will be selected based on their expertise and relevance to the study questions.

Random sampling will be applied to social media users. Stratified sampling will ensure that participants come from a wide demographic segment, including age, gender, and geographic location.

Population Size: A population of approximately 500 participants has been identified, which includes cybersecurity professionals, developers, and experienced social media users. To achieve a robust sample size for analysis, at least 10% (50 respondents) will be targeted to ensure that the data is representative of the broader population.

Data Collection Method: Surveys are the primary method for collecting quantitative data from social media users and IT professionals. The survey instrument will capture information on:

- Authentication methods used
- Perceived effectiveness of authentication methods
- User preference for different authentication methods

The survey will also explore security concerns, blockchain awareness, and experiences with security breaches.

Data Analysis: Closed-ended questions (Likert-Scale and Multiple-Choice) will be used to facilitate statistical analysis, while open-ended questions will capture qualitative insights. Data will be distributed online through email lists and professional networks.

3.2 Data Analysis and Ethical Considerations

After data collection, the data will be sorted and checked for outliers, missing values and abnormalities in preparation for analysis. Descriptive statistics will be the main focus as the researcher will explore the mean, frequencies, percentages and standard deviation of different variables. The findings will then be presented through tables, pie charts, bar graphs and line graphs. However, the researcher will also create cross tabulation tables to examine the relationship between two or more categorical variables. In this case, the study will explore the relationship between familiarity with blockchain technology and perceptions of its security, usability, transparency, trust and privacy. On the other hand, the qualitative data will be analysed using thematic analysis. It is noted that thematic analysis involves coding the data to identify recurring themes and patterns [19]. This will involve a rigorous process of familiarizing the data, generating initial codes, searching for themes, reviewing the themes, and generating a report. The study employs a mixed approach, using a survey and literature review to collect and analyse data pertaining to the research study. Validity and reliability will be ensured through triangulation, standardized procedures, and rigorous data analysis protocols. Ethical considerations will also be achieved by ensuring informed consent, data privacy, and minimization of bias.

4 Results

Each section of this chapter looks into the detailed results of the survey, interpreting and discussing the implications of the findings in the context of established literature.

4.1 Demographic Representation

The below Table 1 highlights the response rate of the people who took part in the survey of the study. As indicated, 100 people out of the 146 people that were recruited for the study, participated in the study. The table further illustrates that 100 respondents constituted to 68.5% while the 46 who failed to participate in research constituted to 31.5% of the target population. According to [20] a response rate of more than 50% is considered to be sufficient to make a credible and conclusive study. As such, a respondent rate of 68.5% was sufficient to collect and analyse quantitative data on the authentication security within popular social media platforms and mobile applications.

Table 1. Response rates

Category	Frequency	Percentage (%)
Responded	100	68.5%
Did not Respond	46	31.5%
Total	146	100%

Gender: Diverse gender distribution is essential in ensuring that a study's responses are not inclined towards a particular gender. In this case, the study's participants included a diverse sample with 51% being male, 36% being female and 9% being non-binary. The remaining 4% preferred not to reveal their genders as indicated by Fig. 3 below. This distribution is relatively balanced and allows for a comprehensive analysis of authentication practices and perceptions across different age groups.

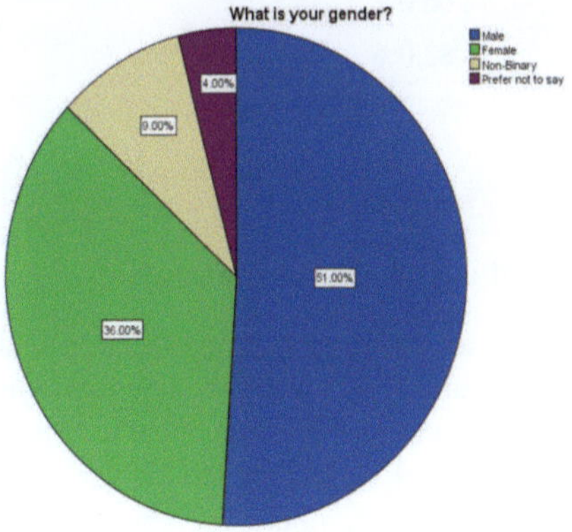

Fig. 3. Gender Representation

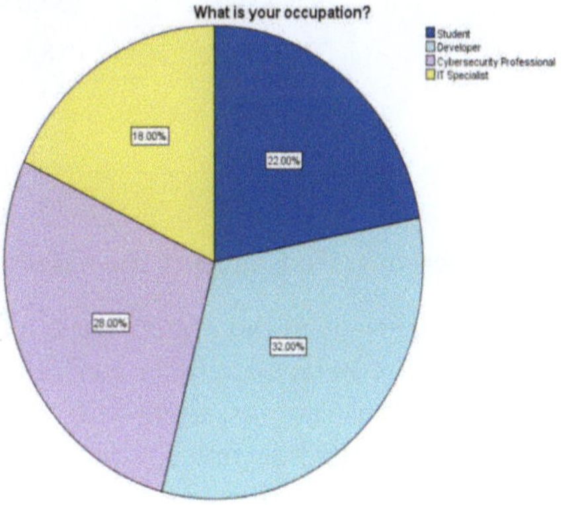

Fig. 4. Gender Representation

Occupation: Figure 4 above indicates the occupational distribution of the respondents. From the pie chart, it is evident that majority of the respondents were developer (32%), followed by cybersecurity professionals (28%), students (22%), and IT specialists (18%). This variety in occupations provided a broad perspective on the authentication methods and security concerns experienced by different segments of the population. For instance, developers and cybersecurity and IT specialists' respondents are more technically savvy and thus would provide a deeper understanding on the practicality and usability of implementing new technologies like blockchain. Lastly, students provide a high level of variability of usability and exposure to the different authentication methods.

Platform Usage: The Table 2 above presents the distribution of the social media platform usage. From the finding, it is evident that Instagram and TikTok are the most commonly used platforms among the respondents with 84% and 82% usage respectively. Facebook and Snapchat are also popular, with usage rates of 68% and 66%. LinkedIn has an 50% usage, while X (Formally Twitter) was used by 53% of the respondents. The data on social media usage reveals key trends that have significant implications for understanding the digital behaviour and preferences of the respondents. To begin with, strong usage of Instagram and TikTok indicates preferences for visual and video-oriented content among the participants. Further, the use of Facebook and LinkedIn highlights the need for ephemeral and professional content respectively.

Table 2. Social Media Platform Usage

	No	Yes
Facebook	32.0%	68.0%
X	47.0%	53.0%
Instagram	16.0%	84.0%
LinkedIn	50.0%	50.0%
Snapchat	34.0%	66.0%
TikTok	18.0%	82.0%

4.2 Objective 1: To Examine the Effective Methods of Authentication

The respondents rated the effectiveness of various authentication methods on a scale of 1 to 5, with 1 being extremely ineffective and 5 being very effective. The following table presents these ratings, along with the mean and standard deviation of each method. Notably, the standard deviation is an essential measure of variability and dispersion in data. In this case, a high standard deviation implies that data points are spread over a wide range. On the other hand, a lower standard deviation implies that the data tend to be close to the mean, implying that the views were unanimous.

From the Table 3 above majority of the respondents opined that passwords were very effective (26.7%). However, 28.3% of the respondents were unsure of whether passwords were indeed effective authentication methods, and this is evident through the mean of 3.37 and standard deviation of 1.25 suggesting variability in opinions. On the other hand,

Table 3. Effectiveness of the Authentication Method

Statement	ENE	NE	U	E	VE	Mean	SD
Passwords	6.1%	21.2%	28.3%	18.2%	26.3%	3.37	1.250
Two-factor Authentication	5.0%	10.0%	21.0%	43.0%	21.0%	3.65	1.077
Biometric Authentication	13.1%	16.2%	25.3%	40.4%	5.1%	3.08	1.140
Single Sign-On	6.0%	18.0%	20.0%	29.0%	27.0%	3.53	1.235

ENE – Extremely Not Effective, NE – Not Effective, U – Unsure, E – Effective, VE – Very Effective

with varied opinions (SD = 1.077), majority of the respondents (43.0%) indicated that two factor authentication was an effective approach (mean = 3.65). Similarly, biometric authentication has a mean effectiveness rating of 3.08 and SD of 1.140 reflecting the missed perceptions of its effectiveness. A notable majority of 40.4% of the respondents opined that biometric authentication was effective. Lastly, with regard to the effectiveness of the Single Sign-On (SSO), the mean effectiveness rating was 3.53, with 56% of the respondents rating it as effective or very effective.

4.3 Objective 2: To Identify Specific Security Risks and Vulnerabilities

The respondents were asked on their opinions regarding the vulnerability of different authentication mechanisms, and the findings were summarized in the stacked bar graph below.

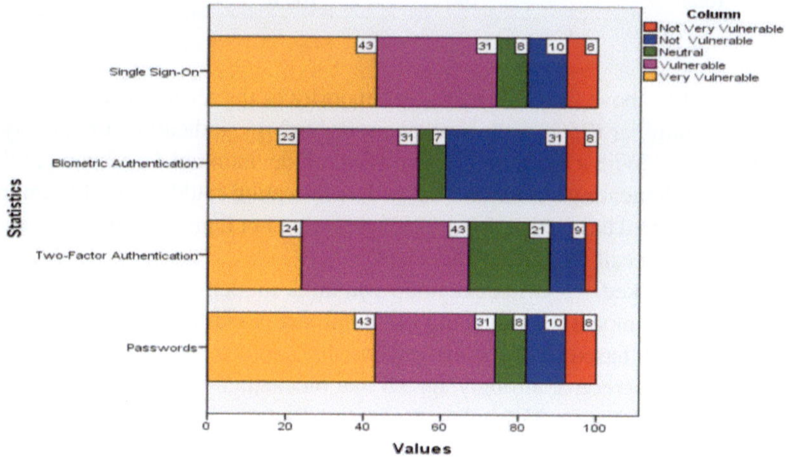

Fig. 5. Vulnerability of Authentication Methods to Security Threats

From the Fig. 5 above, it is evident that passwords were rated as the most vulnerable with a mean of 3.91, with 74% of the respondents considering them vulnerable or very vulnerable. On the other hand, 2FA was also perceived as highly vulnerable (mean = 3.76), with 67% rating it as vulnerable or very vulnerable. Biometric authentication was perceived as moderately vulnerable (mean = 3.30) with mixed views on its security. Lastly, SSO was also viewed as highly vulnerable (mean = 3.91), with 74% considering it vulnerable or very vulnerable. In light of these responses, it can be inferred that high vulnerabilities ratings for passwords and SSO highlight the need for enhanced security measures, while the mixed views on biometrics suggest a need for improvements in biometric technologies.

4.4 Objective 3 and 4: Blockchain Authentication Effectiveness and Implementation

The respondents were asked to express their level of familiarity with blockchain technology and the finding were summarized in Table 4 below.

Table 4. How familiar are you with Blockchain technology?

		Frequency	Percent	Cumulative Percent
Valid	Not Familiar	14	14.0	14.0
	Somewhat Familiar	15	15.0	29.0
	Neutral	4	4.0	33.0
	Familiar	42	42.0	75.0
	Very Familiar	25	25.0	100.0
	Total	100	100.0	

From the table above, it was evident that majority of the respondents were familiar (42%) or very familiar (25%) with blockchain technology, indicating a relatively high level of awareness. With a mean of 3.49 and standard deviation of 1.382, the familiarity suggests that implementing blockchain-based authentication could be feasible among the current well-informed base. Also, for the 29% that were not familiar with the blockchain technology, it was a call for further education.

Further when asked, "To what extent do you think implementing blockchain-based authentication will impact the following on social media platforms?", the respondents had varied opinions based on the statements as illustrated below. To begin with, 65% of the respondents agreed or strongly agreed that blockchain-based authentication will increase user trust. However, 33% either disagree or disagree, indicating mixed opinions. It is noted that blockchain technology's transparency and immutability can enhance trust by ensuring that authentication processes are temper-proof and verifiable [21]. From the insight, users can then have more confidence in the security of their accounts, knowing that their data is protected by advanced cryptographic methods.

When asked whether blockchain-based authentication will enhance data privacy in social media platforms, 82% of respondents agreed while only 14% expressed their disagreement. Lastly, when asked, if blockchain-based authentication will improve overall security in social media platforms, 65% of the respondents agree or strongly agreed, while only 13% were in disagreement. The positive perceptions of blockchain-based authentication in privacy and security suggest that it is a viable strategy for social media platforms. The responses were measured quantitatively through percentages and summarized in the Table 5 below.

Table 5. Implementing blockchain-based authentication will impact social media

Statement	SD	D	N	A	SA	Mean	SD
Blockchain-based authentication will increase user trust in social media platforms.	11.0%	22.0%	2.0%	43.0%	22.0%	3.34	1.343
Blockchain-based authentication will enhance data privacy in social media platforms	4.0%	10.0%	4.0%	57.0%	25.0%	3.89	1.024
Blockchain-based authentication will improve overall security in social media platforms	7.0%	6.0%	18.0%	58.0%	11.0%	3.60	1.006

Each of these methods presented varied challenges and level of vulnerability as indicated by the respondents. Security was also a critical issue with passwords and SSO being perceived as the most vulnerable methods, while phishing and brute attacks being most prevalent. Concerns and insights regarding the use of blockchain-based authentication were also shared by the participants. They noted that improving current methods and addressing challenges related to blockchain adoption can enhance user trust and security on social media platforms.

5 Conclusion

The research confirms the dire need to ensure that secure authentication processes when accessing and using social media platforms. In this case, the study employed a bottom-up approach to collect the views of participants regarding the current methods of authentication used in popular social media platforms, the security risks and vulnerabilities associated with them, blockchain-based authentication and the effectiveness of the emerging technologies in enhancing security and trust. Further, [22] noted that successful research studies may have practical and theoretical implications. In this case, the researcher will also present the implications for both the academic community and industry practitioners.

Current Authentication Methods- The first conclusion that emanated from the study was brought forward by the first research question that sort to establish the main types of authentication methods within the current social media landscape. Inviting responses from a sample of 100 respondents including university students, developers, IT experts, and Cybersecurity professionals, the researcher was able to paint a clear picture of the current landscape of authentication methods. For instance, there was a unanimous agreement on the use of passwords to access social media platforms. In this case, the respondents confirmed to be using passwords even though they also pointed out their concerns regarding its susceptibility to security breaches. Also, it was evident that while two-factor authentication (2FA) provides better security, there were concerns regarding the usability and adaptability of users. The other notable authentication method was biometric, which was seen to offer increased security but faced issues of scepticism and recognition failure.

Security Risks and Vulnerabilities- Considering the existing authentication methods, it was prudent for the research to explore the security risks and vulnerabilities associated with them. The participants expressed their views on what they perceived to be the main security vulnerabilities associated with the existing authentication methods. To begin with, the majority of the respondents pointed out that based on the authentication methods they were using, they were vulnerable to phishing attacks. Another major insight was that for password users (formed by a majority of the participants), there was a high susceptibility to brute force attacks. Lastly, the issue of Man-In-the-Middle Attacks was also reported to be a significant threat to the integrity and safety of user authentication process in social media platforms. The users might have identified the MITM attacks by detecting unusual network activity, unauthorized login locations, and frequent SSL certificate warnings. Addressing the issues of MITM attacks includes regular updates, the use of VPNs, and employing a Multi-Factor Authentication (MFA).

Blockchain-Based Authentication- Emerging technologies were also a critical point of concern for the study. Following the literature review, the researcher was keen to explore the role of blockchain technology in enhancing the integrity and safety of the social media authentication process. In this case, the respondents confirmed to have a reasonable idea to help them make better-informed decisions regarding the use of blockchain technology in enhancing the safety of the authentication process. According to them, blockchain technology is perceived as the more secure and transparent authentication method. However, they also pointed out the complexity of the implementation process and thus, this could be a significant barrier. Lastly, the participants noted that there is a high cost involved in setting up the blockchain technology, and the regulatory hurdles could limit the adoption process.

Effectiveness of Blockchain in Enhancing Trust and Security- The last objective sought to explore the effectiveness of blockchain in enhancing trust and security in social media authentication process. To begin with, the participants noted that blockchain-based authentication could significantly increase user trust in social media platforms. Also, respondents perceived blockchain as a means to significantly enhance data privacy and also the general security of these platforms.

References

1. Bureau of Justice Statistics. Victims of identity theft, 2021 (2023). https://bjs.ojp.gov/library/publications/victims-identity-theft-2021

2. Robinson, R. J. Deciphering generative AI risk trends to integrate compliance as an industry practice. (ICSET-23) (2023). https://ijsem.org/viewabstract.php?id=16553&volume=Volume10&issue=Issue10.
3. Choi, G., Chung, H.: Applying the technology acceptance model to social networking sites (SNS). Int. J. Hum. Comput. Interact. **29**(10), 619–628 (2023). https://doi.org/10.1080/10447318.2012.756333
4. Coelho, K.K., Tristão, E.T., Nogueira, M., Vieira, A.B., Nacif, J.A.: Multimodal biometric authentication method by federated learning. Biomed. Signal Process. Control **85**, 105022 (2023)
5. Robinson, R.J.: Application of artificial intelligence (AI) in higher education (HE) for enhanced learning and employability. In: (AIET Conference 2024), vol. 11, no. 1, pp. 1–6 (2025). https://www.ijlt.org/show-173-1112-1.html
6. Robinson, R.J.: Cloud systems with its security, privacy and trust claims to a sustainable solution. (ICECCME) (2023). https://ieeexplore.ieee.org/document/10252796
7. Creswell, J.W., Creswell, J.D.: Research design: Qualitative, quantitative, and mixed methods approaches (5th ed.). Sage Publications (2017)
8. Das, S., Wang, B., Kim, A., Camp, L.J.: MFA is a necessary chore!: exploring user mental models of multi-factor authentication technologies. In: HICSS, pp. 1–10 (2020)
9. Nakamoto, S.: Bitcoin: A Peer-to-Peer Electronic Cash System (2018). https://bitcoin.org/bitcoin.pdf
10. Robinson, R.J.: Cutting-edge AI techniques for remarkable new product promotions. (ACDSA 2024) (2024). https://ieeexplore.ieee.org/document/10467567
11. Odağ, Ö., Mittelstädt, A.: Mixed methods in research on the psychology of the internet and social media. Qual. Soc. Res. **24**(1) (2023)
12. Peel, K.L.: A beginner's guide to applied educational research using thematic analysis. Pract. Assess. Res. Eval. **25**(1), 2 (2020)
13. Pervin, N., Mokhtar, M.: The interpretivist research paradigm: a subjective notion of a social context. Int. J. Acad. Res. Progressive Educ. Dev. **11**(2), 419–428 (2022)
14. Tata, T.S.: Design of a web-based social network information system. Int. J. Artif. Intell. Res. **6**(1), 310–316 (2023)
15. Taylor, S.J., Bogdan, R., DeVault, M.: Introduction to Qualitative Research Methods: A Guidebook and Resource (4th ed.). Wiley, Hoboken (2015)
16. Wang, C., Wang, Y., Chen, Y., Liu, H., Liu, J.: User authentication on mobile devices: approaches, threats and trends. Comput. Netw. **170**, 107118 (2020)
17. Zyphur, M.J., Pierides, D.C.: Making quantitative research: from positivist dogma to social scientific inquiry. J. Bus. Ethics **167**, 49–62 (2020)
18. Robinson, R.J.: Cybersecurity compliance frameworks-a pragmatic view with an IT outsourcing company case study. Soc. Lens1, 15–25 (2024). https://socialsciencesresearch.com/sl/article/view/6
19. Kumar, S., Lee, J.: Multi-factor authentication: a review of implementation techniques. Int. J. Inf. Secur. **10**(4), 321–336 (2021)
20. Lorenz, T.: Extremely Online: The Untold Story of Fame, Influence, and Power on the Internet. Simon and Schuster. (2023)
21. Ghazi-Tehrani, A.K., Pontell, H.N.: Phishing evolves: analyzing the enduring cybercrime. In: The New Technology of Financial Crime, pp. 35–61. Routledge (2022)
22. Jones, A., Smith, B., Johnson, C.: Password security: a comprehensive analysis of current practices. J. Cybersecur. **3**(2), 145–162 (2018)

Next Generation IoT and Communication Network

A Scalable, Edge Supported Indoor Navigation System

Abinesh Tamizhselvan, R. Kaviswar, Chandana Sandireddy, Shreeya Asawa, and J. Govindarajan(✉)

Department of Computer Science and Engineering, Amrita School of Computing, Amrita Vishwa Vidyapeetham, Coimbatore, Coimbatore, India
{cb.en.u4cse21302,cb.en.u4cse21345,cb.en.u4cse21255, cb.en.u4cse21357}@cb.students.amrita.edu, j_govindarajan@cb.amrita.edu

Abstract. Navigating through large and complex indoor environments such as malls, hospitals, and airports can be challenging due to the unavailability of reliable GPS signals. Most available solutions suffer in terms of scalability, performance, and privacy. The main aim of this work is to develop a scalable, reliable, edge-based indoor navigation system that provides location guidance in real time without relying on cloud-based services. This is done to achieve ease of deployment as well as the efficiency of user navigation. We propose a generalized architecture that is suitable for multi-floor buildings. Our system can be extended to multiple buildings with minimal infrastructure support. The system was successfully deployed and tested in a university building, demonstrating its effectiveness in real-world scenarios.

Keywords: Indoor Navigation System · Edge-Based · Bluetooth Low Energy (BLE) Beacons · A* Algorithm · Real-Time Localization

1 Introduction

Navigation systems have been integrated into life. Their purpose is to assist in the movement of individuals through these various environments. Global Positioning System (GPS) technology is successfully implemented and used for outdoor movements with an inaccuracy of around 4.9 m, given the most favorable conditions [1]. However, GPS signals are interfered with largely by buildings, bridges, and other structures. As such, indoor navigation cannot rely on GPS [2]. These drawbacks have prompted much of the research work into indoor alternative positioning, such as Wi-Fi, Bluetooth Low Energy (BLE), Ultra-Wideband (UWB), and Radio-Frequency Identification (RFID); these are prominent contenders in bringing indoor localizations into focus [2].

One of the most customer-friendly technologies for indoor positioning today is that of BLE beacons, as they consume relatively low power and have a simple setup while aligning with many mobile devices [3]. Undergoing rapid deployment into different venues, such as shopping malls, stadiums, and even airports, these beacons will enable

L. Meng (Ed.): CCCN 2025, SCT, pp. 145–153, 2025.
https://doi.org/10.1007/978-3-032-03632-2_11

enhanced personalized content services to be offered alongside traditional navigation programs [5]. Their adaptive characteristics are also very suitable for smart buildings and smart campuses where effective indoor navigation solutions are needed within spaces like hospitals, universities, and corporate offices [3, 6, 7]. The idea of a smart campus, which combines IoT, AI, and the Internet-fastening future, is gaining weight in the navigation and automatic maintenance arena and in the overall user experience [3–5, 8].

BLE beacons offer many advantages but must integrate suitable positioning techniques for any effective indoor navigation system realized. Most of the existing solutions are either cloud-based or dependent on machine learning- based algorithms; they are non-scalable. Besides, they fall short of a generalized architecture that can apply to any building system. In this regard, our system is scalable, edge-based that provides precise indoor navigation without any hitch. On entering the building, the use of that seamless connection to gain access to an Android application that has an interactive Scalable Vector Graphics (SVG) based map for real-time wayfinding is very interesting. Moreover, we use A* pathfinder to identify the shortest and most efficient route to one's desired destination.

The system has been implemented in our university building, which is a demonstration of the capability of our system to provide effective indoor navigation. This scalable design could be extended to other large indoor spaces, though this approach holds promise for smart campus applications.

2 Literature Survey

Indoor navigation has been a topic of significant research, and many techniques have been developed to improve localization accuracy, reliability, and responsiveness of the system. Although some systems have manifested improvements, most suffer from two kinds of limitations: they either heavily depend on centralized servers with little or no fail-over solutions, or they require internet connection to accomplish their tasks. This section will analyze the related work done in the area of indoor navigation in light of the weaknesses that our proposed system seeks to address.

An indoor navigation system developed on the web and Android was introduced in [9], comprising a modified A* algorithm for pathfinding. It examined human wayfinding behavior through eye-tracking and deep learning models. Lacking in real-time positioning and relying on a web server for functioning, it became unsuitable in contexts characterized by unreliable internet connectivity. Our system is edge-based, allowing for real-time navigation without continuous network access. Moreover, the greater architectural flexibility of our system aligns across other indoor settings, requiring only minor adjustments. A very different pose considered incorporating Augmented Reality (AR) and the semantic web paradigm for a better user experience during navigation [10]. The discussed system comprised four modules: namely, spatial modeling, positioning, content visualization, and data management, with AR-based navigation guidance with virtual overlays. Nonetheless, the positioning methods based on AR are susceptible to lighting conditions and environmental features, rendering them unreliable in dark or visually homogenous environments. Our system, conversely, effectively uses Bluetooth Low Energy (BLE) beacons and edge computing for robust real-time tracking while remaining very accurate in localization irrespective of environmental lighting. Their system operates under a centralized knowledge base, which may include potential points of failure due to network dependency; conversely, our failover mechanism utilizing primary and secondary edge servers ensures uninterrupted operation.

INSUS [11] has looked into implementing Simultaneous Localization and Mapping (SLAM) with AR and QR codes in tracking users. While their system utilized the QR codes for position tracking and Unity for 3D virtual pathfinding, it is also heavily reliant on a central server for computation and data retrieval, which would introduce possible delays and risks when the network functions are disrupted. Our system removes this bottleneck by performing all the navigation data processing on the edge servers and user devices, with reduced latencies and offline functioning provided. Moreover, since the maps are preloaded inside our application, they allow seamless navigation in the absence of any server communications, allowing for a much more robust solution.

In [12], a hybrid BLE and Pedestrian Dead Reckoning (PDR)-based system employing RSS-based multilateration and Kalman filtering was proposed. As the method enhanced its positioning accuracy, it was, however, still dependent upon real-time RSS updates from a central server, giving rise to concerns about network dependency and failures in such systems in the event of connectivity problems. On the contrary, our system minimizes reliance on external servers by using preloaded floor maps and edge computing. Moreover, we add redundancy to our system with an effective multi-server failover mechanism that supports continuous operation even if the primary server fails, thus enhancing the reliability of our system to a level beyond that of present solutions.

An ML-based BLE indoor positioning system was presented in [13], where all KNN, Random Forest, and Extra Trees Classifier (ETC) models were trained to obtain better localization accuracy. It achieved a very high classification accuracy of 95.63 percent. Its reliance on centralized machine learning models and real-time processing of data presents scalability concerns. Our approach, on the other hand, incorporates direct deployment of localization algorithms onto an end user's device, getting rid of the need for a real- time upload of data to a centralized server. Furthermore, our multi-building edge architecture with failover servers guarantees an uninterrupted supply of navigation support, lacking redundancy mechanisms in their case.

What emerges from the above discussion is that existing solutions for indoor navigation remain afflicted by such common problems as centralized process, dependency upon the network, and lack of failover mechanisms. Our proposed edge-supported BLE navigation system looks into resolving these constraints by ensuring real-time, decentralized processing with preloaded maps, edge computing, and a robust failover mechanism. These enhancements improve the reliability, scalability, and real-time performance of the system to a large extent, thus making the proposed solution a better alternative for large-scale indoor applications.

3 Objectives and Proposed Architecture for Indoor Navigation

3.1 Objectives

This section depicts the main considerations in setting up an indoor navigation system. The goal is to develop a strong Bluetooth-based solution with high precision for real-time tracking. The specific objectives thus include:

- Bluetooth-based Indoor Navigation: The implementation of a Bluetooth- based navigation system that provides support for precise real-time navigation and location within an area of the building or room where the GPS signal is very poor or not available.
- Scalability: To design a system that is scalable to any other building with
- minimal infrastructure support.
- Edge Computing Integration: To integrate the edge computing concept to allow local data processing, thus minimizing latency and meeting real-time navigation requirements without heavy dependency on centralized cloud servers.
- User Position Tracking: Allow the users to track their real time position.
- Communication: Establish communication between the mobile and the raspberry pi.

3.2 Proposed Architecture

The proposed architecture (Fig. 1) consists of an edge-based server, Bluetooth Low Energy (BLE) beacons, and an optional central server. As per our architecture, every building requires one edge-based server at the entrance of the building. It serves as a microservices hub to share our application and load map with the user's mobile device upon entry while also providing network connectivity. This avoids the need for an active internet connection from the user device.

For a multi-building scenario, the same setup has to be replicated across n buildings. If there are n buildings, n edge servers are required since one edge server is required per building. Among n edge servers, one acts as a primary server. In our architecture we have also included one optional dedicated server wherever there is a need for some centralized operation. Among n edge servers and one optional server, one functions as a primary server while the others act as secondary servers. If there is a failure in the primary server, the system dynamically makes one of the secondary servers act as a primary server, ensuring uninterrupted operations. The interaction between the primary and secondary server is shown in (Fig. 2).

The Bluetooth beacons are placed strategically in the indoor environment. It serves as a positioning reference points. These beacons transmit the signals which is utilized by the app for the real-time user tracking.

As per our architecture, the shortest path algorithm and localization algorithm are running as part of the Android application.

Since the user's phone is already loaded with a map of the entire building and algorithms are running locally in the user's device, the need for frequent communication with the server is avoided.

The floor map is present in the Android application itself since each floor map size is in kilobytes. This reduces the reliance on network access. Thus, it helps the system function offline. The algorithms run locally in the app.

Moreover, the MQTT broker is chosen according to the device's distance from the local server. When the device is in the vicinity of the local server, the local edge MQTT broker is used; otherwise, the cloud-based MQTT broker is used. Furthermore, the broker is invoked for an update and is not a continuously running process.

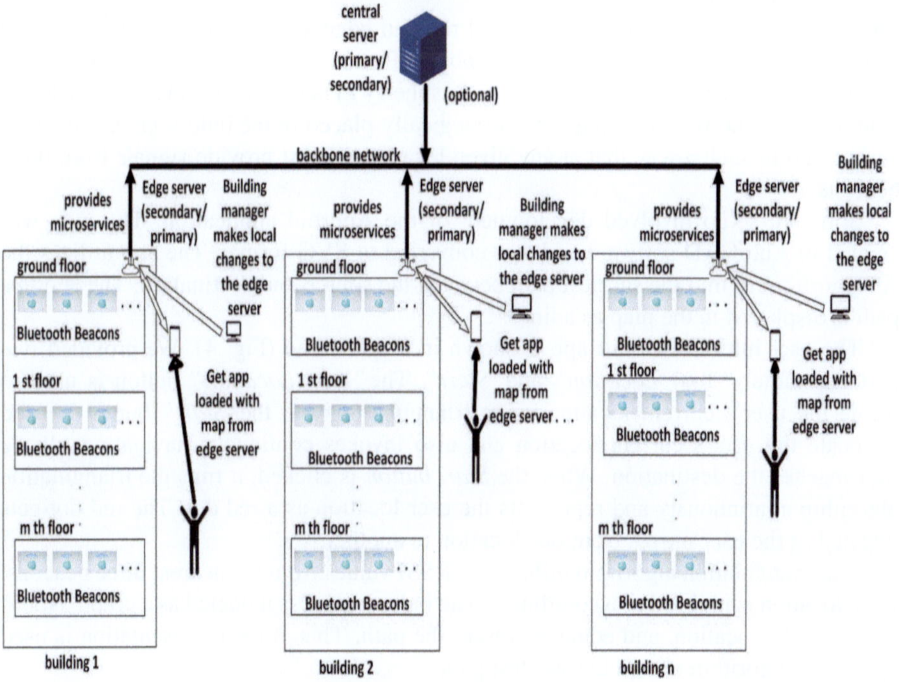

Fig. 1. Proposed System Architecture

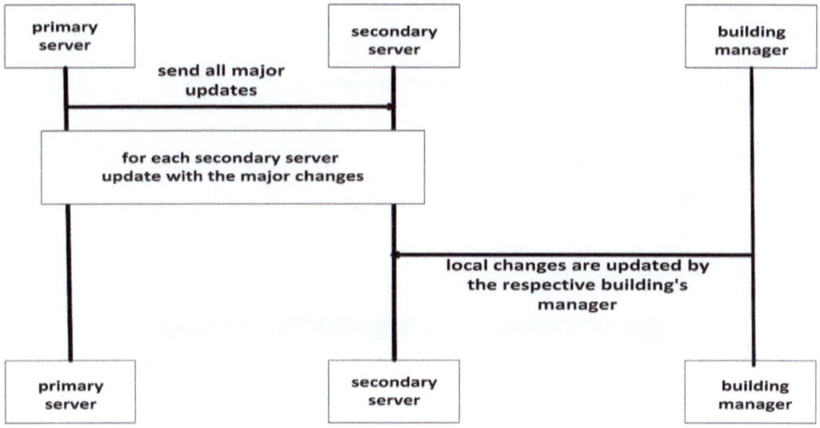

Fig. 2. Interaction between primary and secondary servers.

4 Implementation

For the purpose of validation, the system was demonstrated for one building, which can be extended for multiple buildings.

The implementation of the proposed system involved integrating hardware and software components. The hardware setup involved a Raspberry Pi 5 and BLE beacons. The

Raspberry Pi acts as an access point and runs an edge server to provide the Android application along with the map. The Raspberry Pi was deployed in the entrance of the ground floor. The users can connect to the Raspberry Pi network and get access to download the app. The BLE beacons were strategically placed in the indoor environment. It was placed in such a way that at any given location it must provide signals from three beacons.

Software setup involved development of the Android application. The map was created in AutoCAD software and was converted to SVG format. The app utilizes the A* algorithm to find the shortest path between the source and destination. The shortest path is displayed in the map as a line.

The user interface of our app is shown in (Fig. 3) and (Fig. 4). We provided two buttons, namely*"Your Location"* and*"Start"*. The*"Your Location"* button is used to locate the user's current location using triangulation, and the*"Start"* button is used to locate the user's current location and also invokes continuous navigation till the user reaches the destination. When the *Start button* is clicked, it runs the triangulation algorithm continuously and represents the user location as a red dot. The red dot gets updated as the user moves from one location to another.

The triangulation algorithm utilizes the RSSI values from the nearest three beacons. Each location is mapped as a coordinate. The indoor space is modeled as a graph. Nodes represent the location, and edges represent the path. This graph representation is used by the A* algorithm to find the shortest path.

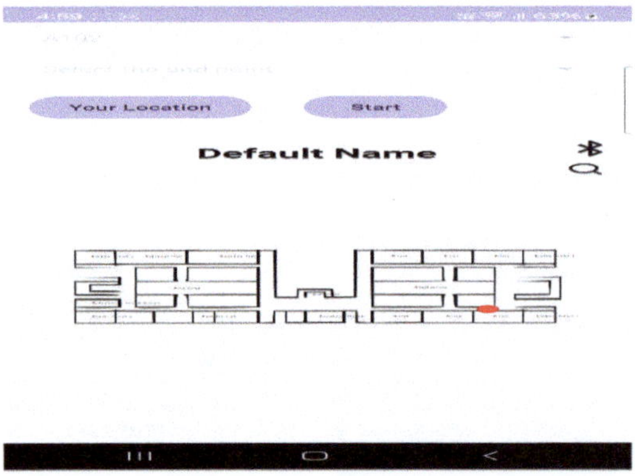

Fig. 3. *"Your Location"* button is clicked

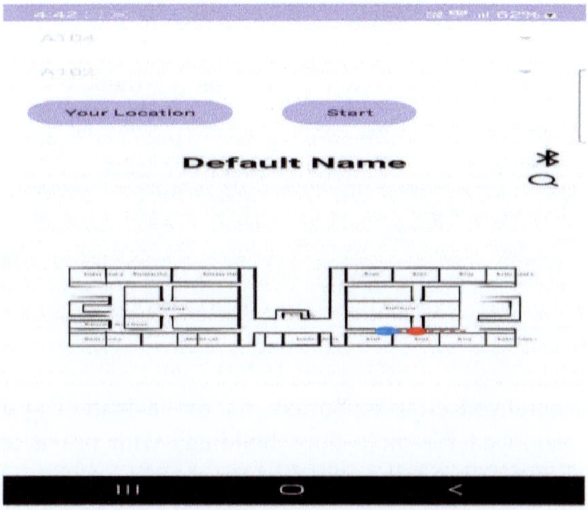

Fig. 4. *"Start"* button is clicked

5 Result Analysis

The localization accuracy, pathfinding efficiency, latency, memory usage, and reliability have been evaluated for the proposed edge-based indoor navigation system and the performance metrics of our system are presented in Table 1. For performance evaluation, 50 navigation trials were performed between the same start and end points. The system is shown to localize at an average error of 1.5 m since RSSI signals fluctuated greatly.

By using the A* pathfinding algorithm, pathfinding was completed in under one second, allowing for rapid route calculation. With position updates every 2–3 s, the system ensured efficient user tracking. Memory consumption was analyzed using Android Studio's profiler, which revealed a shallow size of. 6.4 MB and Retained Memory of 1.36 MB. It Was also Found that no Memory Leaks Occurred and Memory Management Was Efficient.

Data replication was used to enhance system reliability. The secondary server would maintain an operational status for the system in case of a failure of the primary server. The proposed model enhanced stability and reduced latency compared to current GPS and Wi-Fi systems. GPS systems work poorly when indoor blockage occurs, and Wi-Fi-based approaches end up with fluctuating signals and interference. On the contrary, the edge-based technique delivered lower dependency on the network and higher reliability through quicker local processing.

Table 1. Performance Metrics of our proposed system

Metric	Value	Unit
Pathfinding latency	1	seconds
Localization Accuracy	1.5	meters

(*continued*)

Table 1. (*continued*)

Metric	Value	Unit
Position update frequency	2–3	seconds
Shallow size	6.4	MB
Retained memory	1.36	MB

6 Conclusion

We proposed a generalized edge-supported indoor navigation system that can be implemented and deployed for multi-floor buildings with provisions to extend for multi-building scenarios. Our architecture addresses the aspects of scalability and reliability.

Our system employs an edge server (like Raspberry Pi) and Bluetooth beacons for transmission, providing effective real-time indoor navigation. The system has already been put to the test successfully using an Android application that employs an SVG-based floor plan. Within this system, the A* algorithm for path optimization and RSSI-based triangulation for localization are integrated. This means indoor navigation will be both efficacious and timely.

Deployment in our university building showed about 1.5-m mean errors, the main source of which is signal fluctuations in RSSI. However, there were insignificant deviations, and the system was interactive and reliable for navigation support.

Further development will be devoted to making evident our architecture at the scale of an entire university campus, within airports, shopping malls, and hospitals. Improving localization accuracy, using UWB-based localization as part of our future work, will also be included. With these advancements, the system has the potential to serve as a reliable and efficient indoor navigation solution across various applications.

References

1. Hoffman-Wellenhof, H., Lichtenegger, B., Collins, J.: Global Positioning System: Theory and Practice. Springer, New York (2001)
2. Zafari, F., Papapanagiotou, I., Devetsikiotis, M., Hacker, T.J.: Enhancing the accuracy of iBeacons for indoor proximity-based services. In: IEEE International Conference on Communications, pp. 1–7. IEEE (2017)
3. Booranawong, A., Sengchuai, K., Jindapetch, N.: Implementation and test of an RSSI-based indoor target localization system: human movement effects on the accuracy. Measurement **133**, 370–382 (2019)
4. Alghamdi, A., Shetty, S.: Survey toward a smart campus using the Internet of Things. In: 2016 IEEE 4th International Conference on Future Internet of Things and Cloud (FiCloud), pp. 235–239. IEEE (2016)
5. Villegas-Ch, W., Palacios-Pacheco, X., Lujan-Mora, S.: Application of a smart city model to a traditional university campus with a big data architecture: a sustainable smart campus. Sustainability, **11** (2019)

6. Nair, A.P., Ajay, V., Manoj, A.K., Nihal, M.P., Asha, C.A.: Indoor navigation system using Wi-Fi RSSI. In: 2024 International Conference on Circuit Power and Computing Technologies (ICCPCT). IEEE (2024). https://doi.org/10.1109/ICCPCT61902.2024.10672863
7. Nagarajan, B., et al.: Localization and indoor navigation for visually impaired using bluetooth low energy. Smart systems and IoT: innovations in computing. In: Proceeding of SSIC 2019. Springer, Singapore (2020)
8. Sai Pranavdhara Reddy, N., et al.: Enhancing indoor navigation efficiency with WiFi signal mapping. In: 2024 15th International Conference on Computing Communication and Networking Technologies (ICCCNT). IEEE (2024). https://doi.org/10.1109/ICCCNT61001.2024.10724456
9. Ludwig, B., Donabauer, G., Ramsauer, D., al Subari, K.: URWalking: indoor navigation for research and daily use. KI Ku¨nstliche Intelligenz **37**, 83–90 (2023). https://doi.org/10.1007/s13218-022-00795-1
10. Rubio-Sandoval, J.I., et al.: An indoor navigation methodology for mobile devices by integrating augmented reality and semantic web. Sensors **21**(16), 5435 (2021). https://doi.org/10.3390/s21165435
11. Fajrianti, E.D., et al.: INSUS: indoor navigation system using unity and smartphone for user ambulation assistance. Information **14**(7), 359 (2023). https://doi.org/10.3390/info14070359
12. Dinh, T.-M.T., Duong, N.-S., Nguyen, Q.-T.: Developing a novel realtime indoor positioning system based on BLE beacons and smartphone sensors. IEEE Sens. J. **21**(20), 23055–23068 (2021)
13. Maaloul, K., Lejdel, B., Clementini, E., Abdelhamid, N.M.: Bluetooth beacons-based indoor positioning in shopping malls using machine learning. Bull. Electr. Eng. Inform. **12**(1), 911–921 (2023). https://doi.org/10.11591/eei.v12i2.4200

A Systematic Review Using MPNet-Based Correlated Topic Model for IoT Security

Ching-Hsue Cheng(✉)

National Yunlin University of Science and Technology, Douliou, Yunlin 64002, Taiwan
chcheng@yuntech.edu.tw

Abstract. IoT security issues are a huge burden for users who lack relevant knowledge or skills when they are faced with processing large amounts of information in a short period of time. In order to allow the public to effectively understand IoT security, this study collected 2018–2023 IoT-related literature from the Web of Science as a research dataset to extract key information and conduct subsequent analysis for the public to use easily. Feature extraction is one of the important steps in this study. This study proposed a MPNet_CTM model to build topic model, which applied the masked and permuted net (MPNet) to process IoT security articles and used correlated topic model (CTM) to process topic model. In terms of topic numbers, this study applied correlated topic model and the consistency score to build the best topic numbers. In verifying effectiveness of topic model, there are five claasifiers used for classification and evaluation, including decision tree, random forest, XGBoost, Adaboost, and CNN. After BERT and MPNet natural language processing, this study compares BERT + LDA, BERT + CTM, MPNet + LDA, and MPNet + CTM. The results show that the MPNet + CTM based on XGBoost classifier is the best performance. Finally, this study uses VOSviewer to visually present the connection strength of the extracted words, and to demonstrate the keyword strength and connection of IoT information security. The results can help IoT users better understand the trends and challenges of IoT information security.

Keywords: IoT security · Natural language processing (NLP) · Masked and permuted net (MPNet) · Correlated topic model (CTM) · Knowledge graph

1 Introduction

The Internet of Things (IoT) is an innovative technology that integrates various processes of identification, sensing, networking, and computing to promote the combination of physical and virtual objects. The rapid development and application of IoT has brought new challenges, especially in security and privacy. According to Forbes [1], the number of malware incidents on IoT devices has increased from 813 million in 2018 to 2.9 billion in 2019. These security risks are not limited to a single device, they affect entire systems and networks. According to the 2023 IoT security landscape report [2], there have been as many as 3.6 billion security incidents based on approximately 120 million

L. Meng (Ed.): CCCN 2025, SCT, pp. 154–165, 2025.
https://doi.org/10.1007/978-3-032-03632-2_12

IoT devices worldwide. Smart homes are one of the most important applications of IoT, there are 2.6 million smart homes in the world, and the networks of these homes suffer an average of 8 attacks on devices every 24 h [2]. These facts show that as the number of IoT devices grows exponentially, so does the number of attacks on these devices.

Natural language processing (NLP) has succussed application in many fields, one of the famous NLP is the bidirectional encoder representations from transformers (BERT) proposed by Google [3]. BERT is a pre-training technology for natural language processing, it uses a bidirectional design to consider the relationship between words in the text and enhance the model. The other worth paying attention is the masked and permuted language model (MPNet [4]). MPNet using masking and displacement technology can better meet the challenges encountered by BERT, it makes the results of topic modeling more reliable and distinctive [4]. Based on NLP, many researchers have proposed tools and methods to analyze text information, especially text mining, which can extract key words from a large amount of text [5]. Text mining aims to extract words with specific meanings in the entire text collection, and then treat these words as feature input to perform key extraction and analysis. One of the tools for key extraction is topic modeling [5], which can analyze unstructured text based on unsupervised learning in machine learning without manual labeling.

The most widely used topic modeling technology is latent Dirichlet allocation (LDA [6]). LDA is an unsupervised machine learning technology that uses a three-layer Bayesian probability model to identify hidden topic information in large-scale texts [6], and a text represents a probability distribution composed of several topics. Each topic consists of some words and a different probability distribution over the words. To improve LDA issue, Blei and Lafferty [7] proposed the correlated topic model (CTM), which is an extension of LDA that allows the occurrence of topics to be correlated to relax the assumptions made by LDA. CTM uses the logistic normal distribution to model topic correlation instead of the Dirichlet distribution in LDA, which explicitly allows documents to have associated topics. Therefore, when the topics of many articles are related, we can use the Logistic-Normal model to build Correlated LDA.

This paper proposed a MPNet-based CTM model to build the best topic numbers of IoT security. This study first collected academic literature on IoT and IoT security from the Web of Science, and we used BERT and MPNet natural language processing to extract features of the collected texts. Further, the LDA and CTM are applied to build topic model, and the consistency score is used to find the optimal number of topics. And then the optimal topics are used as input variables for classification evaluation to verify the performance of topic modeling.

2 Related Work

This section conducts a automating literature review on IoT information security, text mining, and the classification technology used in this study.

2.1 IoT Security

An IoT is defined as a network interconnection model supported by a technology stack composed of multiple technologies, which makes the data flow and interaction between sensors and virtual cloud services closer and smoother. Technology stack is the integration of multiple technologies, such as sensing, communication, data processing, and data analysis. The combined effect of these technologies enables anyone and anything to be connected anytime and anywhere, which can interact with IoT systems anywhere [8]. The IoT first started with a three-layer architecture, which is perception layer, network layer, and application layer. The application layer processes and analyzes data and then adds it to various applications, such as smart home control, health monitoring, and logistics tracking [9]. Now, IoT has been extended to the four-layer architecture and the five-layer architecture of the network and application layers.

From the term IoT was coined in 1999, the security issues of the IoT have not received enough attention. Until 2011, more and more smart watches, smart home appliances, and devices are connected to the Internet. Further, the increase in data leaks, privacy violations, and cyber-attacks have attracted public and corporate attention [9]. Due to increasingly severe security threats, many organizations have formulated security standards and regulations for IoT. Based on the evolution of attack methods, IoT security is also constantly improving, including the use of more complex encryption technologies, security protocol updates, and the use of artificial intelligence (AI) and machine learning technologies to predict and prevent security threats [10]. A detailed review of the security challenges and threat sources in the IoT applications were presented by Hassija et al. [11].

2.2 Natural Language Processing

NLP is addressed to define appropriate and efficient token, sentence, and document representations [12]. In recent NLP, BERT developed by Google in 2018 is an unsupervised pre-trained language representation model to better understand the semantics of user search sentences [3]. BERT uses the bidirectional encoder of the transformer architecture to learn the deep contextual relationship of the language. It can consider the context of the single word in the sentence at the same time, and it is very powerful in understanding language.

BERT ignores the semantic connections between obscured words, resulting in the loss of some contextual information. Therefore, the generalized autoregressive language model (XLNet) was extended and the permutation language model (PLM) method was used to improve the limitations of BERT. In addition, XLNet did not utilize the complete position information of the sentence, which resulted in position differences between pre-training and fine-tuning [4]. To improve this problem, MPNet inherited the advantages of BERT and XLNet to improve their limitations [4]. MPNet is a natural language pre-training model proposed by Microsoft. It combines BERT's masked language model and XLNet's PLM to better handle the challenges of vocabulary order and ambiguity, making NLP results more reliable and distinct [4]. The performance of MPNet is significantly better than MLM and PLM, and compared with BERT, XLNet, and RoBERTa under the same model settings [4]. The applications of MPNet are also quite diverse. Searching the past literature, the authors have not seen research on MPNet applied to IoT.

2.3 Topic Modeling

Topic modeling uses unsupervised learning algorithms on large amounts of text to produce a set of term summaries, which can generate common keywords or phrases from text datasets and group these words into multiple topics [13]. The topic modeling used in this study is introduced below.

Latent Dirichlet Allocation (LDA). LDA is the most widely used topic modeling technology [14] (Kherwa & Bansal, 2020). In LDA, each document is assumed to consist of a set of topics that follow a Dirichlet probability distribution [6]. For each document, LDA first uses the Dirichlet distribution α to initialize the topic distribution (θ) of the document. i.e., $\theta_{i=1...D} = Dirichlet_D(\alpha)$. . LDA also uses the Dirichlet distribution β for each topic to initialize the probability that each word belongs to the topic. Next, the LDA model first randomly selects a topic from the topic distribution (θ_i) of the document, and a word is randomly selected from the word distribution (φ) corresponding to the selected topic. This process is repeated until all words in the document are generated.

Correlated Topic Model (CTM). Due to LDA assumes topic independence, which means that the topic distribution in a document is not affected by the topic distribution of other documents. However, topic distribution may not always follow an independent distribution, thereby the CTM is proposed by Blei and Lafferty [7]. CTM utilizes μ and Σ to replace LDA's α to reflect that the model captures the correlation between topics. This distribution contains measures of covariance between topics, which allows for a more realistic description of how the presence of one potential topic may be related to that of another potential topic [14].

3 Proposed Method

To enable public users to effectively understand IoT security, we collect the great well-known web of science journal articles to establish an IoT security knowledge map for the public can quickly understand IoT security. Based on the aim of visualized knowledge map, this study proposes an MPNet-based CTM to handle IoT security topics, and we use BERT and MPNet natural language processing to extract features of the collected text. Further, the LDA and CTM models are applied to build a topic model, and the consistency score is used to find the optimal number of topics. The optimal topics are then used as input variables for classification evaluation to verify the performance of topic modeling. The proposed research procedure with four steps and the detailed steps are described as follows.

Step 1. Collect Data. *This step uses the keyword IoT and "IoT security" on the topic field to search and restrict English articles from the Web of Science database between 2018 and 2023.* For collected all articles, each article includes title, abstract, author keywords, keywords plus, publication year, and research areas. The research topics of the collected articles cover 254 research areas from Web of Science, and each article is assigned to at least one research area. Based on each article may be assigned to multiple research areas, in order to avoid conflicts in overlapping research areas, only a single research area is remained. The top 3 research areas have total 14542 articles on the IoT from 2018 to 2023. Further, the IoT security has total 4350 articles from 2018 to 2023 in the top 3 research areas. Therefore, the experimental dataset contained 10192 articles on the IoT and 4350 articles on IoT security and we named "IoT_security" dataset.

Step 2. Preprocess Text. This step uses the MPNet pre-training model for the collected articles, compared to traditional text pre-processing models, the MPNet can better extract contextual relationships to deal with the challenges of vocabulary order and ambiguity, making the results of topic modeling more accurate. After performing word tokenizing, we will perform lemmatization, restore each word to its basic form, and create unigram, bigram, and trigram to generate vocabulary combinations that often appear in text. Based on application view, we briefly introduce MPNet. MPNet applies the Transformer architecture and incorporates masked and permuted language modeling to pre-train objective for improving performance. This following is its main steps:

1. Permute and mask tokens: After tokenizing all texts. Randomly permute the token sequence of a sentence. In the predicted part, we mask tokens, and obtain the new input tokens and the corresponding position sequence.
2. Model output dependency with two-stream self-attention: we follow permuted language modeling to adopt two-stream self-attention to autoregressively predict the tokens
3. Bidirectional modeling to extract the representations: In the non-predicted part, we use bidirectional modeling to extract the representations.
4. Position compensation: We use position compensation to ensure the model can see the full sentence and reduce the input inconsistency between pre-training and fine-tuning.

Step 3. Build Topic Model by CTM. This step constructs the topic model by using CTM. First, this study uses the results of NLP preprocessing to create text vectors based on bag of words. For the parameter adjustment of the CTM topic model, alpha and beta are hyperparameters that affect topic sparsity, and this study uses the default value without adjustment. Finally, the probability of each text in each topic is extracted by the CTM topic model as a feature for subsequent treatments.The reasons for using CTM in this study are: (1) CTM can capture the correlation between topics as latent semantics [7]; (2) CTM replaces LDA with a logistic normal distribution, which contains the covariance between topics; (3) it allows for a more realistic description of the fact that the presence of one latent topic may be related to the presence of another latent topic.

In evaluating the best number of topics, this study used the consistency score Cv because it has the best consistency score after comparing multiple datasets [15]. The Cv combines indirect cosine similarity, normalized pointwise mutual information (NPMI), and boolean sliding window. It creates a length N for each topic word word vectors, and operates by comparing the similarity between high-ranking words within a single topic [15].

Step 4. Classification and Evaluation. In order to verify the effectiveness of the best CTM topic numbers for IoT security, this study uses five kinds of classifiers: Decision Tree (DT [16]), Random Forest (RF [17]), Adaptive Boosting (AdaBoost [18]), eXtreme Gradient Boosting (XGBoost [19]), and Convolutional neural network (CNN [20]) for model training and verification. The five classifiers were used to perform verification based on 10-fold cross-validation and the constructed the optimal number of topics is classification features. The classification label has 10192 articles on the IoT and 4350 articles on IoT security.

Previous research results show that accuracy, precision, recall, F1, and AUC are the most commonly considered performance indicators [21, 22]. Further, AUC is used to compare different models but the premise is that the records of data in different classes should be roughly balanced. The larger area indicates that the model has better matching and performance [23].

4 Experiments and Results

4.1 Experimental Environment and Parameter Settings

The experimental environment is a computer with RTX2080 Ti graphics card and RAM 64G, and the Python 3.8.3 is installed on the Windows 10 system; further, the Visual Studio code is used for programming and compilation in this study. We use the Gensim and Sklearn for text preprocessing and topic pattern modeling, they are free python module. This study uses the DT, RF, XGBoost, AdaBoost and CNN classifiers to classify the IoT and IoT security classes and ten-fold cross-validation is used to evaluate the performance of the constructed model. The parameter settings including the five classifiers, LDA, and CTM are shown in Table 1.

4.2 Experimental Results

This section compared the proposed model with the listed models and visulalized topic models. First, we apply BERT and MPNet natural language to process the collected IoT_security dataset. Second, this study uses LDA and CTM to build the topic model and determine the best topic numbers. And then, we use the best topic numbers as classification attributes to classify IoT and IoT security for verifying the performance of the built topic model. The experimental results are described as follows.

Determine the Best Topic Numbers. This study sets 200 topics as the maximum number of topics, and we calculate the Cv consistency score of every 5 topics successively to determine the optimal number of topics. The higher coherence scores indicate that the topic is more coherent and easier to explain. In addition, the MPNet is an expansion of BERT, and the CTM is an improvement of LDA; therefore, the BERT and LDA are included in the experimental comparison. Finally, the all combination of the NLP and topic models are named BERT_CTM, BERT_LDA, MPNet_CTM, and MPNet_LDA. First, this study plots the determining graphs of these four combination models based on the abscissa of topic number vs Cv ordinater, where we only show the determined graphs of MPNet_CTM, as shown in Fig. 1. From the experimental analysis, the highest coherence score is the best topic numbers, and their results are shown Table 2.

Among four combining models, the highest consistency score is the MPNet_CTM model, as shown in Table 2, which presents that the MPNet_CTM is more fitable than the other combination models for topic modelling in the collected IoT_security dataset.

Table 1. The parameter settings of the five classifiers.

Algorithm	Parameter	Reference
DT	criterion = gini, max_depth = None, max_features = None, max_leaf_nodes = None, min_impurity_decrease = 0.0, min_samples_leaf = 1, min_samples_split = 2, splitter = best, ccp_alpha = 0.0	Quinlan [16]
RF	criterion = gini,max_depth = None, max_features = auto, max_leaf_nodes = None, min_impurity_decrease = 0.0,oob_score = False,random_state = None,verbose = 0,warm_start = Fasle,ccp_alpha = 0.0	Breiman [17]
XGBoost	colsample_bynode = None,colsample bytree = None,gamma = None,n _estimators = 100,n_jobs = None,num_parallel_tree = None,random_state = None,reg_alpha = None,reg_lambda = None,scale_pos_weight = None,subsample = None, tree_method = None,validate_parameters = None,verbosity = None	Chen & Guestrin [19]
AdaBoost	algorithm = SAMME.R,base_estimator = None,learning_rate = 1.0,n_estimators = 50,random_state = None	Freund & Schapire [18]
CNN	Opt = adam, learning_rate = 0.001	Yamashita [20]
CTM	distributed = False, chunksize = 2000, passes = 1, update_every = 1, alpha = 0.1, eta = 0.01, decay = 0.1, offset = 1.0, eval_every = 10, iterations = 50, gamma_threshold = 0.001, minimum_probability = 0.01, minimum_phi_value = 0.01, per_word_topics = True, callbacks = None, dtype = < class 'numpy.float32' >	Blei & Lafferty [7]
LDA	distributed = False, chunksize = 2000, passes = 1, update_every = 1, alpha = 'symmetric', eta = None, decay = 0.5, offset = 1.0, eval_every = 10, iterations = 50, gamma_threshold = 0.001, minimum_probability = 0.01, minimum_phi_value = 0.01, per_word_topics = False, callbacks = None, dtype = < class 'numpy.float32' >	Blei et al. [6]

Regardless of BERT and MPNet word processing, the number of words of the CTM is less than the LDA topic modelling. Because there are correlations between CTM topics, and some words may be shared in multiple related topics, which results in a relatively small number of words extracted in CTM. However, the LDA topic independence prefers to assign words to a single topic; hence, the number of words extracted in LDA will be larger. These findings represent that the CTM is more coherent and easier to explain than LDA model.

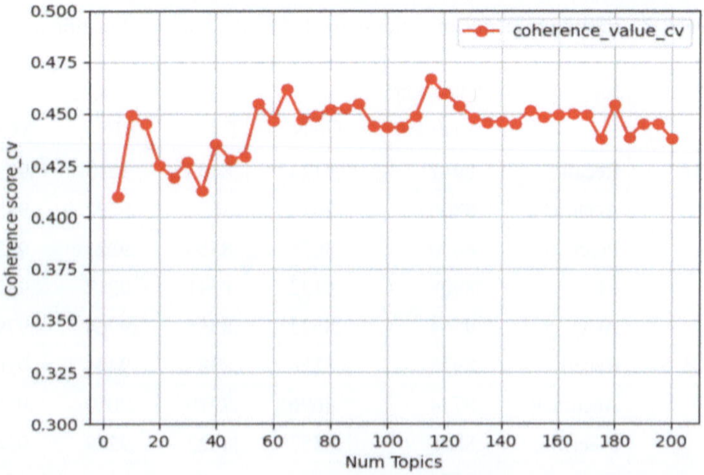

Fig. 1. Determining the best topic numbers graph of the MPNet_CTM for topic number vs Cv.

Table 2. The best topic numbers, highest Cv, and word numbers for four combination models.

Model	Best topic numbers	Highest Cv score	# word
BERT_LDA	40	0.4336	139,972
BERT_CTM	155	0.4535	68,211
MPNet_LDA	60	0.4232	149,614
MPNet_CTM	115	0.4667	66,493

Verify The Performance of Topics Extracted. After topics extracted by BERT_CTM, BERT_LDA, MPNet_CTM, and MPNet_LDA models for the collected IoT_security dataset, we use these topics extracted of each model as classification features to evaluate the effectiveness of the extracted topics. The verification is based on ten-fold cross-validation to classify the IoT and IoT Security classes. In performance metrics, we use accuracy, recall, precision, F1, and AUC to evaluate their performance, and. Finally, the comparative results of the extracted topics by four combination models are shown in Table 3. From Table 3, the extracted topics by the MPNet_CTM model perform better than other combination modes in accuracy, precision, F1 score, and AUC metrics under the XGBoost classifier. This means that the MPNet_CTM model has the best extracted topics and explanation capabilities.

Visulized Words Extracted by MPNet_CTM. From experimental results (2), the MPNet_CTM model was the best extracted topics and explainable ability. Furthermore, among the four combination models, the number of words extracted by MPNet_CTM is the smallest with 66,493 words. Therefore, we used the VOSviewer [24] version 1.6.20 to visualize words extracted by MAPNet_CTM. The VOSviewer is a tool specifically

Table 3. Comparative results of topics extracted by the four combination models.

Model	Metric	Classifier				
		Adaboost	CNN	DT	RF	XGBoost
BERT_CTM	Recall	.8947	.8124	.8403	.9213	.9239
	Accuracy	.8866	.8018	.8674	.9222	.9439
	Precision	.8890	.8056	.8550	.9221	.9328
	F1	.8908	.8132	.8541	.9217	.9329
	AUC	.9564	.8525	.8528	.9764	.9798
BERT_CTM	Recall	.8975	.8156	.8387	**.9404**	.9263
	Accuracy	.8734	.8696	.8710	.9050	.9259
	Precision	.8828	.8357	.8523	.9228	.9243
	F1	.8870	.8603	.8507	.9251	.9260
	AUC	.9542	.9068	.8495	.9770	.9771
MPNet_LDA	Recall	.8933	.8559	.8357	.9215	.9236
	Accuracy	.8912	.7722	.8674	.9228	.9438
	Precision	.8907	.7961	.8503	.9219	.9325
	F1	.8920	.8810	.8501	.9223	.9326
	AUC	.9567	.9045	.8496	.9757	.9797
MPNet_*CTM*	Recall	.9073	.8556	.8683	.9242	.9241
	Accuracy	.8985	.8752	.8907	.9297	**.9508**
	Precision	.9016	.8616	.8800	.9279	**.9364**
	F1	.9031	.8775	.8791	.9267	**.9361**
	AUC	.9631	.9394	.8795	.9792	**.9816**

designed to visualize bibliometric networks, which can establish and display the co-occurrence relationships between important terms in the literature, thereby providing insights into the overall structure and key themes of the research field. In a visual knowledge graph, a node represents each keyword, and its size represents the importance of the node and its influence in the knowledge graph; the connecting line between nodes represents the association between the two nodes, and its thickness indicates the strength of the association. These VOSviewer symbols allow us to graphically understand the connections between documents, so as to better explore and analyze the development trends in the research field.

First of all, the overall 6-year academic articles of the IoT_security dataset have previously obtained the best number of 115 topics and 66493 keywords from MPNet_CTM model. Too many topics and keywords will result in the inability to clearly present important knowledge content. Therefore, this study selected the top 10 most important topics and the top 20 keywords for each topic, then we used VOSviewer to present the knowledge graph, as shown in Fig. 2. From Fig. 2, we observe some important issues as follows.

(1) In the blue block, the keyword "Security" and "Scheme" have become the most prominent clusters, which shows that increasing popularity and application of IoT devices need to focus on security and scheme. Security cluster includes how to protect IoT devices and networks from unauthorized access and attacks, as well as how to ensure data privacy and integrity. Scheme cluster refers to various strategies and solutions proposed to solve these security problems, such as encryption technology, security protocols and identity authentication systems.
(2) In the green block, "Service" is a prominent cluster, which highlights the importance of providing seamless services across devices, platforms, and systems in the IoT environment. This involves communication, data exchange, automated services and enhanced functions between IoT devices, all of which improve efficiency and user experience.
(3) The "Power" cluster in the red block reflects the focus on energy consumption and performance of IoT devices. As the number of devices increases, how to effectively manage energy consumption, improve energy efficiency, and ensure continued operation has become a key issue.
(4) Finally, the "Protocol" in the purple block and the "Accuracy" in the yellow block reveal the important role of the protocol in ensuring the performance and accuracy of the IoT system. "Protocol" is related to the efficiency and security of data transmission; while "Accuracy" is related to the accuracy of data processing and analysis, which is crucial for IoT applications based on data-driven decision-making.

5 Conclusion

This study has proposed MPNet_CTM model build a topic model for IoT security. That is, the text preprocessing is based on MPNet NLP, which can handle word order and ambiguity better than the traditional NLP that only relies on word frequency. And then, we use CTM topic model to build the topic model. From experimental results, the MPNet_CTM model has the highest Cv consistency score. Further, we apply five classifier to verify the performance of topics extracted by the four different combination models, and the result shows that the MPNet_CTM model perform better than other combination modes in accuracy, precision, F1 score, and AUC metrics under the XGBoost classifier. According to the experimental results, the contributions of this study are as follows.

1. Proposed MPNet_CTM model to establish an IoT security topic model;
2. Apply the highest consistency score Cv to get the best topic numbers and use five classifiers to verify the effectiveness of the topics extracted by MPNet_CTM; and
3. Use VOSviewer to visualize the knowledge graph for IoT security.

In future work, the advancement of technology has made the architecture of the Internet of Things increasingly complex, and a more in-depth study can be conducted on the trends of IoT security articles. Further, there are AI tools of literature analysis can be applied to this research field.

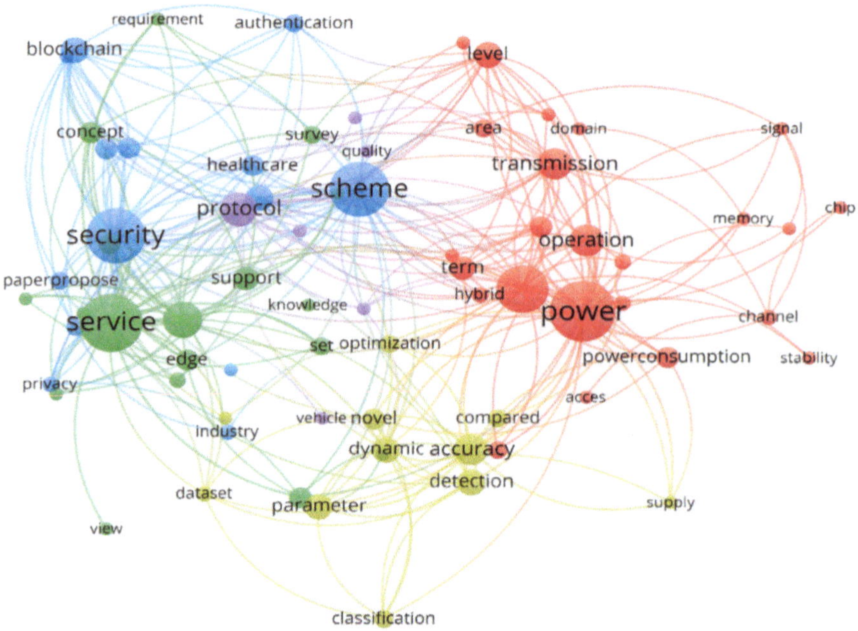

Fig. 2. Knowledge graph of IoT and security articles.

References

1. Forbes: Cyberattacks On IOT Devices Surge 300% In 2019, 'Measured In Billions', Report Claims. https://www.forbes.com/sites/zakdoffman/2019/09/14/dangerous-cyberattacks-on-iot-devices-up-300-in-2019-now-rampant-report-claims/. Accessed 29 Apr 2023
2. Bitdefender: The 2023 IoT Security Landscape Report. https://www.bitdefender.com/files/News/CaseStudies/study/429/2023-IoT-Security-Landscape-Report.pdf. Accessed 07 June 2025
3. Devlin, J., Chang, M.-W., Lee, K., Toutanova, K.: BERT: pre-training of deep bidirectional transformers for language understanding. arXiv preprint arXiv:1810.04805 (2018)
4. Song, K., Tan, X., Qin, T., Lu, J., Liu, T.-Y.: MPNet: masked and permuted pre-training for language understanding. In: Advances in Neural Information Processing Systems, vol. 33, pp. 16857–16867 (2020)
5. Zhang, R., et al.: A literature-mining method of integrating text and table extraction for materials science publications. Comput. Mater. Sci. **230**, 112441 (2023). https://doi.org/10.1016/j.commatsci.2023.112441
6. Blei, D.M., Ng, A.Y., Jordan, M.I.: Latent dirichlet allocation. J. Mach. Learn. Res. **3**(Jan), 993–1022 (2003)
7. Blei, D.M., Lafferty, J.D.: A correlated topic model of science. Ann. Appl. Stat. **1**(1), 17–35 (2007). https://doi.org/10.1214/07-aoas114
8. Čolaković, A., Hadžialić, M.: Internet of things (IoT): a review of enabling technologies, challenges, and open research issues. Comput. Netw. **144**, 17–39 (2018). https://doi.org/10.1016/j.comnet.2018.07.017
9. Alshohoumi, F., Sarrab, M., AlHamadani, A., Al-Abri, D.: Systematic review of existing IoT architectures: security and privacy issues and concerns. Int. J. Adv. Comput. Sci. Appl. **10**(7), 232–251 (2019)

10. Chaganti, R., Ravi, V., Pham, T.D.: Deep learning based cross architecture internet of things malware detection and classification. Comput. Secur. **120**, 102779 (2022). https://doi.org/10.1016/j.cose.2022.102779
11. Hassija, V., Chamola, V., Saxena, V., Jain, D., Goyal, P., Sikdar, B.: A survey on IoT security: application areas, security threats, and solution architectures. IEEE Access **7**, 82721–82743 (2019)
12. Lauriola, I., Lavelli, A., Aiolli, F.: An introduction to deep learning in natural language processing: models, techniques, and tools. Neurocomputing **470**, 443–456 (2022)
13. Jurafsky, D., Martin, J.: Speech and language processing: an introduction to natural language processing, Computational Linguistics, and Speech Recognition, 3rd edn. https://web.stanford.edu/~jurafsky/slp3/. Accessed 07 June 2025
14. Kherwa, P., Bansal, P.: Topic modeling: a comprehensive review. EAI Endorsed Trans. Scalable Inf. Syst. **7**(24), 2 (2020). https://doi.org/10.4108/eai.13-7-2018.159623
15. Röder, M., Both, A., Hinneburg, A.: Exploring the space of topic coherence measures. In: Proceedings of the 8th International Conference on Web Search and Data Mining (WSDM 2015), Shanghai, China, pp. 399–408 (2015)
16. Quinlan, J.R.: Induction of decision trees. Mach. Learn. **1**, 81–106 (1986)
17. Breiman, L.: Random forests. Mach. Learn. **45**, 5–32 (2001)
18. Freund, Y., Schapire, R.E.: A decision-theoretic generalization of on-line learning and an application to boosting. J. Comput. Syst. Sci. **55**(1), 119–139 (1997). https://doi.org/10.1006/jcss.1997.1504
19. Chen, T., Guestrin, C.: XGBoost: a scalable tree boosting system. In: Proceedings of the 22nd ACM SIGKDD International Conference on Knowledge Discovery and Data Mining (KDD 2016), San Francisco, pp. 785–794 (2016). https://doi.org/10.1145/2939672.2939785
20. Yamashita, R., Nishio, M., Do, R.K.G., Togashi, K.: Convolutional neural networks: an overview and application in radiology. Insights Imaging **9**, 611–629 (2018)
21. Zain, Z.M., Sakri, S., Ismail, N.H.A.: Application of deep learning in software defect prediction: systematic literature review and meta-analysis. Inf. Softw. Technol. **158**, 107175 (2023)
22. Johnson, J.M., Khoshgoftaar, T.M.: Survey on deep learning with class imbalance. J. Big Data **6**(1), 1–54 (2019)
23. Fawcett, T.: An introduction to ROC analysis. Pattern Recogn. Lett. **27**(8), 861–874 (2006)
24. Perianes-Rodriguez, A., Waltman, L., van Eck, N.J.: Constructing bibliometric networks: a comparison between full and fractional counting. J. Informet. **10**, 1178–1195 (2016). https://doi.org/10.1016/j.joi.2016.10.006

Machine Learning-Driven Wireless Optimization: Improving Network Performance and ISP Selection

Yugavashini Chandran[1], Rafidah Md Noor[1(✉)], Muhammad Umair Munir[1], Ismail Ahmedy[1], Bryan Raj Peter Jabaraj[1], and Norazlina Khamis[2]

[1] Faculty of Computer Science and Information Technology, Universiti Malaya, Kuala Lumpur, Malaysia
{fidah,ismailahmedy}@um.edu.my, bryanraj15@um.edu.mym

[2] Faculty of Computing and Informatics, Universiti Malaysia Sabah Kota Kinabalu, Kota Kinabalu, Sabah, Malaysia
norazlina@ums.edu.my

Abstract. Wireless communication remains crucial as the deployment of 5G networks expands and 4G continues to be essential. The integration of emerging technologies into corporate processes has intensified competition and created new opportunities for service providers. Businesses are shifting focus from acquiring to retaining customers due to the growing dependence on mobile devices across individual, corporate, and governmental sectors. Ensuring proper resource allocation, network design, and Quality of Service (QoS) is vital for implementing advanced wireless technologies globally. Among QoS metrics, network throughput is critical for user satisfaction. However, validating the authenticity and reliability of throughput measurements at vendor endpoints is equally important. Despite contributions from multiple telecommunications providers, QoS remains inconsistent in some regions. This study evaluates the performance of 2G, 3G, 4G, and 5G technologies, identifying which providers deliver the best service in specific areas. The research uses machine learning to predict the most suitable Internet Service Provider (ISP) based on parameters like download/upload speeds and latency. This research is significant for analyzing and identifying the best network provider that supports optimal download and upload speeds and latency in specific areas. Furthermore, it can help network providers analyze and target required regions based on user requirements and use cases, ultimately delivering the best network experience for everyone.

Keywords: Wireless communication · Artificial Intelligence · Network frequency · Emerging Technologies

1 Introduction

The mobile internet era has significantly transformed consumer behavior, with escalating demands for internet accessibility and a wide range of service providers offering mobile communication services. The rapid evolution of mobile networks, wireless sensor networks, broadband services, and the transition from 4G to 5G technologies has positioned

L. Meng (Ed.): CCCN 2025, SCT, pp. 166–179, 2025.
https://doi.org/10.1007/978-3-032-03632-2_13

these systems as essential components of modern communication infrastructure [1, 2]. Wireless technologies have become necessary across several industries, including banking, agriculture, logistics, healthcare, and education, underpinning critical functions in the global economy [3]. Researchers have extensively analyzed how mobile data influences consumer and societal behaviors. For instance, studies using mobile phone data have explored population dynamics, while investigations into mobile app usage and academic performance have uncovered complex patterns, including inverted U-shaped relationships [4]. Modern applications, such as artificial intelligence (AI) and the Internet of Things (IoT), require higher bandwidth to maintain the Quality of Service (QoS) in communication networks [5, 6].

Telecommunication companies have acknowledged the role of wireless communication in enhancing consumer satisfaction and actively investing in improving network experiences. Reliable and fast connectivity is now a basic requirement for modern living and economic development. Enhanced telecommunications infrastructure has been linked to reduced costs associated with economic activities such as information retrieval, transportation, and market exploration [7, 8].

In Canada, leading telecommunications providers like Bell, Rogers, and Telus, which offer services from 2G to 5G networks, dominate the wireless communication landscape. These providers are continually pressured to enhance their networks to improve speed, coverage, and cost-efficiency, as advancements in network generations significantly affect user experiences. Moreover, as the service provided by mobile operators increased, options to choose the most suitable provider become more complicated as different users have different network frequency requirements. Likewise, telecommunication providers are challenged by a lack of precise insights into user preferences and high-performing network frequencies across regions. To bridge these gaps, this study employs the K-Nearest Neighbor (KNN) algorithm to analyze the optimal network frequencies and identify the top Internet Service Providers (ISPs) in specific areas. This analysis not only assists consumers in selecting reliable networks but also helps ISPs in enhancing network stability, targeting underserved areas, and increasing profitability.

The paper is organized into several sections, including a discussion of related works in Sect. 2, methodology in Sect. 3, results presented in Sect. 4, and the conclusion in Sect. 5.

2 Related Work

2.1 Machine Learning

Machine learning algorithms are widely employed to develop robust and effective categorization models. The selection of the dataset is crucial for analyzing mobile network performance predictions, as the precision of the model's predictions heavily depends on it. Gupta et al. [9] emphasize that the dataset must be representative of real-world scenarios and include a diverse range of variables and parameters relevant to mobile network performance analysis, enhancing the robustness and generalizability of the predictive model. Wahid has noted that the volume of data used for prediction affects accuracy; as data volume increases, the machine's ability to learn and train also improves, which also leads to enhancing the model accuracy [10, 11]. According to Elmasry, splitting the dataset into 90% for training and 10% for testing yields optimal accuracy [12].

A critical consideration in evaluating model accuracy is the abundance of features and parameters integrated into the prediction. However, predicting mobile network performance does not necessitate the use of every feature in the dataset. For computing a more accurate model, researchers must consider feature importance and visualize classifier rules. Typically, most models should account for classification, as reliance on classification can introduce biases in selecting features crucial for analyzing the performance of the most effective ISPs. Numerous systems and methods are available for assessing broadband network experiences, though they are not universally accessible. Broadband speed is the most crucial and commonly used measure for evaluating network quality [13]. It is also the most significant factor in determining the overall effectiveness of a broadband network. Consequently, having a tool, algorithm, or platform to assess broadband network speed is essential for network providing firms.

2.2 KNN Algorithm Compared to Other Existing Machine Learning Algorithms

The K-Nearest Neighbor (KNN) algorithm is one of the simplest methods for addressing classification challenges and offers several advantages over other data mining techniques, often yielding competitive results. The KNN algorithm serves as the foundation for developing an adaptable, transparent, straightforward, and easy-to-understand method. As noted by researchers [14], the KNN approach has been evaluated and found capable of addressing certain limitations inherent in other existing algorithms. According to [15], the KNN classification model classifies an individual data point by utilizing the k nearest neighbors. Initially, the model identifies the closest neighbors by calculating the distance between data points based on their attributes and weights. Then, using the selected nearest neighbors, the model categorizes a new example into a class. The dominant class among the k nearest neighbors dictates the predicted class based on proximity [16]. KNN is advantageous in several ways, such as it addresses the scalability issue common to many other data mining techniques through its ability to manage training data too large to fit in memory. Similarly, the implementation of KNN is straightforward as it does not require prior knowledge about the data distribution and relies on simple Euclidean distance to measure similarities between training and test tuples. KNN also reduces the error rate caused by inaccuracies in the assumptions made by other techniques, such as the Naïve Bayesian classification, which often presumes class conditional independence and requires probability data that may not be readily available.

3 Methodology

This study employs the OSEMN framework to model and evaluate the KNN algorithm for a comparative analysis of network frequencies and Internet Service Providers (ISPs) [17]. The methodology consists of five primary steps aimed at predicting our target network frequency and network service provider. The initial phase involves "Obtaining" data from verified sources. The subsequent "Scrub" phase involves cleaning and processing the data into a format suitable for machine learning training. This phase is critical as noisy data can significantly impede the accuracy of predictions. The "Explore" phase involves performing Exploratory Data Analysis (EDA) to analyze data behaviors, trends, and patterns (Fig. 1).

Fig. 1. Methodology

First, data is gathered from relevant sources than the data undergoes cleaning process to remove noise and unwanted elements, and categorical data is converted to numerical format to meet the requirements of KNN classification. Addresses, including city, region, and country, are constructed to filter data specific to specific areas. Additionally, network types such as 5G, 4G, and 3G are categorized based on average download speeds. After preprocessing and cleaning, Exploratory Data Analysis (EDA) is conducted to examine the data before initiating machine learning activities. Important features impacting prediction results are identified using a correlation heatmap.

Next, data modeling begins. The dataset is split into various training and testing ratios, such as 60/40, 70/30, 80/20, and 90/10, to ensure reliable testing and evaluation. The training dataset builds a KNN model for classifying Internet Service Providers (ISPs) and network frequencies. These systematic steps ensure data readiness, insightful analysis, and robust modeling for accurate classification. Further details on each step are outlined in the subsequent sections.

3.1 Obtain - Data Collection

To facilitate informed decision-making in areas such as internet connectivity, policy, development, and education, Ookla LLC provides "Ookla open datasets" at no cost. This study primarily utilizes Ookla's mobile internet performance statistics [18]. The "Global Fixed Broadband and Mobile Network Maps" dataset was employed for this feasibility assessment. Ookla is recognized as a global leader in internet speed testing and analysis. These speed tests, characterized as client-based measures, assess the condition of broadband internet access. While alternative methods for measuring internet speed exist, Ookla is considered the most reliable source for evaluating internet speed. The crowdsourced network tests offer a comprehensive view of network performance, with data specifically collected, processed, and analyzed for Canada. The dataset used spans one year, divided into three quarters of 2023, comprising 9,322 records across 12 attributes, detailed in Table 1 below.

Table 1. Table captions should be placed above the tables.

Field Name	Type	Description
avg_d_kbps	integer	The average download speed of all tests.
avg_u_kbps	integer	The average upload speed of all tests.
avg_lat_ms	integer	The average latency of all tests
avg_lat_down_ms	integer	The average download latency under load of all tests
avg_lat_up_ms	integer	The average upload latency under load of all tests
Tests	integer	The number of tests
Devices	integer	The number of unique devices
Quadkey	text	Representing the tile
Tile	alphanumeric	Polygon of coordinate of an area
tile_x	numeric	X coordinate of tile's centroid
tile_y	numeric	Y coordinate of tile's centroid

3.2 Scrub-Data Preprocessing

The study begins with cleaning the data to ensure a concise, noise-free, and consistent dataset. Issues such as missing data in critical location coordinates, specifically in attributes like tile_x and tile_y, are addressed by removing entries lacking complete data, as imputation could lead to distortions in geographic accuracy. Address fields, including city, state, and country, are derived from these coordinates to ensure the dataset pertains specifically to Canada. Frequency attributes are delineated to classify network types—5G, 4G, 3G, and 2G—according to established frequency ranges. Null values in numerical records are replaced with "0", and object records are labeled as "Unknown" to minimize biases in the dataset. Removing or imputing these values is avoided to maintain data integrity. These steps are essential to maintain the integrity of the data. Categorical attributes, such as network frequency and ISP, are transformed into numerical formats to ensure compatibility with machine learning algorithms, thus supporting the accuracy of subsequent predictions and classifications.

3.3 Explore - Exploratory Data Analysis (EDA)

Exploratory Data Analysis (EDA) was conducted to deepen our understanding of the dataset and to uncover potential insights, a crucial part of our study. We calculated the central tendencies of each attribute, assessed correlation coefficients between the attributes and the output variable, and identified the nature of the data—whether discrete or continuous. These steps are critical as they inform the settings used in the machine learning models and help predict the expected outcomes. This analytical process ensures that the models are appropriately calibrated and that their predictions are reliable.

3.4 Heatmap (Correlation Coefficiency Analysis)

A correlation heatmap is a visual tool that illustrates the relationships and dependencies among various variables in a dataset. It represents the degree of correlation between each pair of variables through a color-coded matrix, facilitating a quick and comprehensive understanding of data patterns and associations. In this matrix, a value of −1 indicates no correlation, whereas a value of 1 indicates perfect correlation. Correlation heatmaps are crucial for identifying variables that may potentially result in multicollinearity, which could compromise the integrity of the predictive model.

Correlation matrix calculations are applicable only to numerical data. Prior to calculating the correlation matrix, features that are irrelevant, such as IDs and categorical features, are removed from the dataset.

Based on the correlation heatmap shown in Fig. 2, we observed that:

1. Network frequency has a 72% positive correlation with tile_x and a 51% positive correlation with tile_y, suggesting that each coordinate may define a network frequency.
2. Both average download speed and average upload speed show a positive correlation of 59%.
3. Average upload and download latencies, measured in milliseconds, exhibit a positive correlation of 37%.

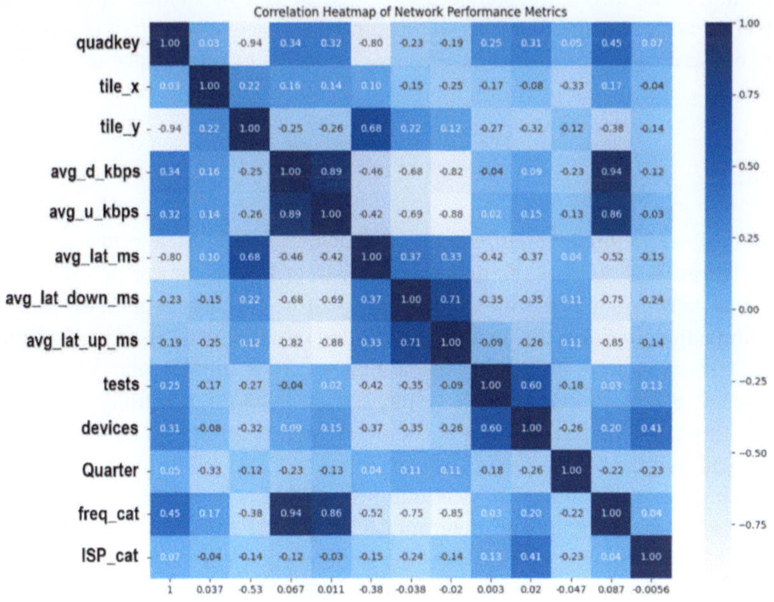

Fig. 2. Correlation Heatmap

3.5 Modeling

The K-Nearest Neighbors (KNN) model is a classification technique employing k nearest neighbors to categorize a given data point. Initially, the KNN model identifies the closest neighbors by assessing their proximity based on their attributes and respective weights. Then, the model assigns a classification to a new instance by analyzing these identified nearest neighbors. Classification can be determined by either assigning weights based on the distance between the new instance and its closest neighbors or selecting the class to which the majority of the neighbors belong. This method relies on measuring the similarity between incoming data and previously collected training data to predict a class. In implementing the KNN algorithm, the model calculates individual distances using the Euclidean metric. The anticipated class is determined by the most prevalent class among the subset of k nearest neighbors that are closest in distance.

Different train-test split ratios are utilized to evaluate the model's accuracy and determine the lowest false positive rate. The dataset is split into training and testing sets to ensure data scalability without leakage. To analyze findings thoroughly, and to prevent overfitting the results, the data splitting ratios are as follows:

The data splitting ratio is as follows:

- 60% of training data and 40% of testing data
- 70% of training data and 30% of testing data
- 80% of training data and 20% of testing data
- 90% of training data and 10% of testing data

As part of the data preparation process, standardization involves centering the input (features) by removing the mean and then dividing by the standard deviation, resulting in data with a mean of 0 and a standard deviation of 1. This standardization is crucial when dealing with machine learning algorithms that may be sensitive to variations in the scale of the data's characteristics, ensuring consistent performance across different data metrics.

4 Results

Based on the correlation heatmap, we selected several attributes demonstrating significant correlations with network frequency to construct and refine our predictive model. Experiments were conducted with varying values of *K*, contingent upon different train-test splitting ratios. Initially, the dataset was divided into four sets based on the following ratios: 60% training and 40% testing as the first set, 70% training and 30% testing for the second set, 80% training and 20% testing for the third set, and finally 90% training and 10% testing for the last set. The accuracy of each splitting ratio was meticulously compared and validated.

For each dataset configuration, maximum K-neighbors have been explored by plotting the validation error curve to determine the optimal *K* value that yields the best accuracy. The F1-Score was employed as the primary metric for selecting the optimal *K* nearest neighbors in our analysis. The decision to utilize the F1-Score stems from its ability to provide a balanced evaluation of model performance by incorporating both recall and precision components. This is particularly advantageous as the F1-Score addresses both aspects, whereas accuracy metrics might only focus on recall, overlooking precision. Consequently, the F1-Score is widely favored in machine learning research and practice for performance assessment. We applied consistent methodologies and algorithms across all splitting ratios and will present the results of the best-performing splitting ratio alongside a comparative evaluation of the other ratios.

4.1 Network Frequency

In this study, we analyzed two datasets: one processed with SMOTE to balance the data and another without SMOTE, presenting inherent imbalances. The experiments conducted to determine the accuracy of network frequency classification revealed that the highest accuracy, 95%, was achieved with a data split of 80/20 and K-neighbors set to 4 for the dataset without SMOTE. Conversely, for the SMOTE-enhanced dataset, the highest accuracy of 95% was observed with a splitting ratio of 90/10 and K-neighbors set to 1.

For the non-SMOTE dataset, the optimal F1-score and an accuracy of 95.77% were achieved with $k = 6$ (refer to Fig. 3). Similarly, for the SMOTE dataset, the peak accuracy of 98.15% was recorded at k=3 (refer to Fig. 4). In both instances, a training and testing split of 90% and 10%, respectively, were utilized to ascertain the optimal K-neighbors, ensuring consistent results across tests. These findings underscore the efficacy of the K-NN algorithm in accurately classifying network frequencies in datasets, regardless of data balancing.

Further analysis of the imbalanced dataset focused on the classification accuracy across different network frequencies. Based on the network frequency data available for Canada, classifications were segmented into four labels: 5G, 4G, 3G, and 2G. As illustrated in Fig. 5, the class labeled 5G demonstrated superior classification accuracy compared to other classes, achieving 95% accuracy. Additionally, the metrics for class 5G showed a precision of 97%, a recall of 99%, and an F1-score of 98%.

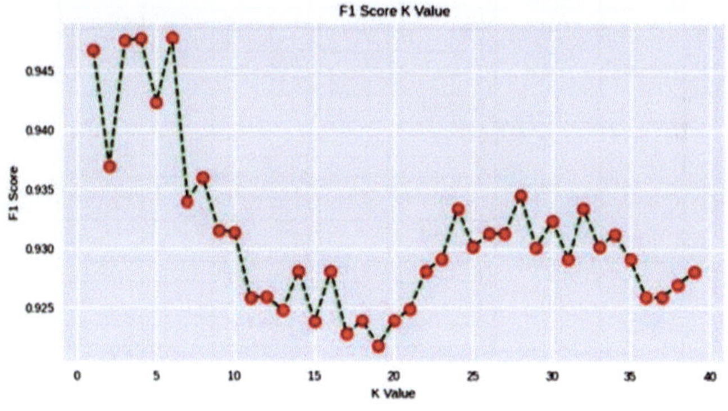

Fig. 3. F1-Score by K-Neighbor Without SMOTE

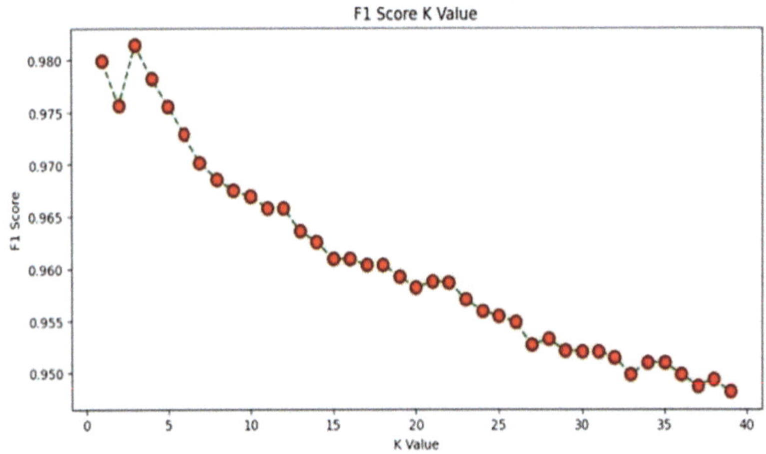

Fig. 4. F1-Score by K-Neighbor With SMOTE

Meanwhile, for the balanced dataset, we can validate the classification of each class to determine the highest accuracy among the network frequencies. Based on the class available for network frequency of Canada data, we could observe that the class has been classified into 4 labels which are 5G, 4G, 3G and 2G. Figure 6 shows that the accuracy of class 2G is classified correctly compared to the rest of the class with 98% accuracy. The accuracy, precision, recall and F1-Score of class 2G are 98%, 98%, 100% and 100%, respectively. This has led to the overfitting of the model prediction.

Splitting ratio	K-Neighbour	Class	Accuracy	Precision	Recall	F1 Score
60/40	3	5G	0.94	0.94	0.97	0.95
		4G	0.94	0.96	0.93	0.94
		3G	0.94	0.89	0.96	0.92
		2G	0.94	0.71	0.40	0.51
		Weighted Avg.	0.94	0.94	0.94	0.94
70/30	4	5G	0.94	0.96	0.95	0.96
		4G	0.94	0.96	0.93	0.94
		3G	0.94	0.87	0.98	0.92
		2G	0.94	0.80	0.40	0.53
		Weighted Avg.	0.94	0.94	0.94	0.94
80/20	4	5G	0.95	0.96	0.96	0.96
		4G	0.95	0.97	0.94	0.95
		3G	0.95	0.88	0.99	0.93
		2G	0.95	0.89	0.47	0.62
		Weighted Avg.	0.95	0.95	0.95	0.95
90/10	6	5G	0.95	0.97	0.99	0.98
		4G	0.95	0.96	0.94	0.95
		3G	0.95	0.87	0.96	0.92
		2G	0.95	0.75	0.27	0.40
		Weighted Avg.	0.95	0.95	0.95	0.95

Fig. 5. Accuracy, precision, recall and F1-Score of Network frequency based on classification without SMOTE.

Splitting ratio	K-Neighbour	Class	Accuracy	Precision	Recall	F1 Score
60/40	1	5G	0.97	0.97	0.97	0.97
		4G	0.97	0.95	0.93	0.94
		3G	0.97	0.96	0.98	0.97
		2G	0.97	1.00	1.00	1.00
		Weighted Avg.	0.97	0.94	0.94	0.94
70/30	3	5G	0.97	0.97	0.98	0.97
		4G	0.97	0.97	0.93	0.95
		3G	0.97	0.96	0.99	0.98
		2G	0.97	0.99	1.00	1.00
		Weighted Avg.	0.97	0.97	0.97	0.97
80/20	3	5G	0.98	0.98	0.98	0.98
		4G	0.98	0.98	0.93	0.95
		3G	0.98	0.96	0.99	0.98
		2G	0.98	0.99	1.00	1.00
		Weighted Avg.	0.98	0.98	0.98	0.98
90/10	3	5G	0.98	0.98	0.98	0.98
		4G	0.98	0.98	0.95	0.96
		3G	0.98	0.97	0.99	0.98
		2G	0.98	0.99	1.00	1.00
		Weighted Avg.	0.98	0.98	0.98	0.98

Fig. 6. Accuracy, precision, recall and F1-Score of Network frequency based on classification with SMOTE.

4.2 Internet Service Provider (ISP)

The experiment evaluating ISP accuracy on imbalanced and balanced datasets highlighted the performance of the KNN algorithm. For the imbalanced dataset, KNN without a specified k-neighbor achieved 79.19% accuracy (Fig. 7). However, identifying the optimal k-neighbor for all splitting ratios revealed that the 70/30 train-test split yielded the best accuracy of 80.12%. Using the balanced dataset, the initial accuracy was 79.93%, while the 90/10 split provided the highest accuracy of 82.67%.

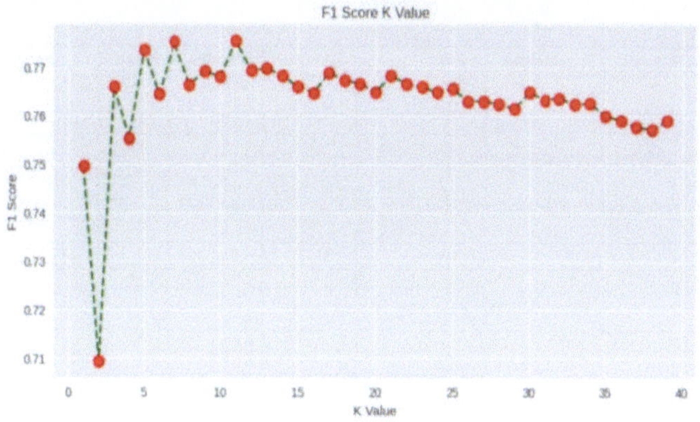

Fig. 7. K-neighbour based on F1-Score without SMOTE.

For the imbalanced dataset, the F1-score analysis identified k = 11 as the optimal neighbor, achieving 77.55% accuracy with a 70/30 split. Similarly, for the balanced dataset, k = 1 provided 80.19% accuracy with a 90/10 split (Fig. 8). The argmax()

function demonstrated the KNN algorithm's effectiveness in improving accuracy and precision across varying train-test splits.

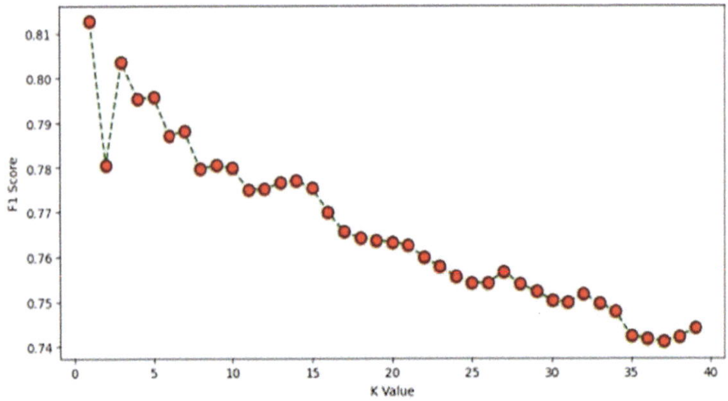

Fig. 8. K-neighbor based on F1-Score with SMOTE.

Figure 9 shows that the accuracy of class ISP Roger is classified correctly compared to the rest of the class for the 80% accuracy. The accuracy, precision, recall and F1-Score of class Roger is 80%, 80%, 96% and 87%, respectively. Next, we validated the classification of each class to determine the highest accuracy among the ISPs available, like Telus, Roger, and Bell, for the balanced data. Figure 10 shows that the accuracy of class ISP Roger is classified correctly compared to the rest of the class for the 83% accuracy achieved with the balanced dataset that was generated using an oversampling technique with SMOTE. The accuracy, precision, recall and F1-Score of class Roger are 83%, 86%, 87% and 87%, respectively.

Splitting ratio	K-Neighbour	Class	Accuracy	Precision	Recall	F1 Score
60/40	9	Telus	0.80	0.59	0.15	0.24
		Roger	0.80	0.81	0.95	0.87
		Bell	0.80	0.81	0.82	0.81
		Weighted Avg.	0.80	0.78	0.80	0.78
70/30	11	Telus	0.80	0.57	0.15	0.24
		Roger	0.80	0.80	0.96	0.87
		Bell	0.80	0.82	0.82	0.82
		Weighted Avg.	0.80	0.78	0.80	0.78
80/20	5	Telus	0.80	0.58	0.24	0.34
		Roger	0.80	0.82	0.94	0.88
		Bell	0.80	0.80	0.82	0.81
		Weighted Avg.	0.80	0.78	0.80	0.78
90/10	7	Telus	0.79	0.62	0.20	0.30
		Roger	0.79	0.82	0.94	0.88
		Bell	0.79	0.78	0.83	0.80
		Weighted Avg.	0.79	0.78	0.79	0.77

Fig. 9. Accuracy, precision, recall and F1-Score of ISP based on classification without SMOTE.

Splitting ratio	K-Neighbour	Class	Accuracy	Precision	Recall	F1 Score
60/40	1	Telus	0.80	0.79	0.70	0.74
		Roger	0.80	0.85	0.84	0.84
		Bell	0.80	0.79	0.70	0.74
		Weighted Avg.	0.80	0.80	0.80	0.80
70/30	1	Telus	0.81	0.78	0.89	0.83
		Roger	0.81	0.86	0.84	0.85
		Bell	0.81	0.80	0.71	0.75
		Weighted Avg.	0.81	0.81	0.81	0.81
80/20	1	Telus	0.82	0.79	0.89	0.84
		Roger	0.82	0.86	0.85	0.85
		Bell	0.82	0.81	0.71	0.76
		Weighted Avg.	0.82	0.82	0.82	0.82
90/10	1	Telus	0.83	0.78	0.91	0.84
		Roger	0.83	0.86	0.87	0.87
		Bell	0.83	0.85	0.71	0.77
		Weighted Avg.	0.83	0.83	0.83	0.82

Fig. 10. Accuracy, precision, recall and F1-Score of ISP based on classification with SMOTE.

5 Discussion and Future Directions

This study evaluated the K-NN algorithm's efficacy in predicting network frequencies and Internet Service Providers (ISPs) by analyzing accuracy, precision, recall, and F1-score across various data splits. For network frequency, the highest accuracy of 95% was achieved using a 90% training and 10% testing split, with 5G predictions being notably accurate among the 5G, 4G, 3G, and 2G classes. ISP classification attained an accuracy of 80% with 70% training and 30% testing split, with Rogers emerging as the most accurately predicted ISP. The K-NN algorithm demonstrated robust performance, reinforcing its reliability for predicting high-performing networks like 5G and ISPs such as Rogers. While the results offer valuable insights into network performance and ISP prediction within the Canadian telecommunications context, extrapolating these findings to other countries necessitates careful consideration due to several factors:

- Telecommunication infrastructure and adoption rates of technologies like 5G differ significantly worldwide.
- Geographic and demographic differences could lead to varying network performance outcomes.
- Market dominance by providers like Rogers or Telus in Canada may not align with provider dynamics in other markets.
- The dataset's quality, size, and diversity directly affect model performance; applying the model to another country may require significant modifications.
- Consumer behaviors and economic factors vary globally, influencing network and ISP performance.

Building upon these limitations, we propose several directions for future research. To improve the generalizability of our findings, future work should conduct cross-country or cross-region comparisons using publicly available datasets from diverse geographical areas. This approach would help identify common patterns and differences in network performance and ISP selection, enabling the development of more universally applicable models. Moreover, future research should perform additional statistical tests and domain-specific analyses for advanced feature analysis to better understand the causal relationships between features. Our correlation analysis revealed that network frequency has a 72% positive correlation with tile_x and a 51% positive correlation with tile_y.

These relationships warrant deeper investigation into how geographical factors such as terrain and population density influence network performance. Furthermore, advanced feature engineering techniques could be applied to create new features or transform existing ones, potentially improving model predictive power and providing more interpretable results for telecommunications stakeholders.

While our study utilized KNN with promising results, implementing and comparing the performance of additional state-of-the-art machine learning algorithms would provide more robust insights. Future work should implement other state-of-the-art machine learning algorithms alongside KNN, reporting results using consistent performance metrics and data splits. This comparison should evaluate prediction accuracy, computational efficiency, and interpretability, which are crucial factors for the practical deployment of such systems in real-world telecommunications environments.

6 Conclusion

This study demonstrated the effectiveness of the K-NN algorithm for predicting optimal network frequencies and ISPs based on performance metrics and geographical data. Through methodical testing across multiple data splits, classification accuracies of 95% for network frequency and 80% for ISP selection were achieved. The KNN algorithm's performance validates its application for network optimization and provider selection in telecommunications. In this study, our proposed approach offers dual benefits: consumers can make informed decisions about service providers based on location-specific predictions, while telecommunications companies can identify areas for network enhancement and strategic development. The model's ability to accurately classify 5G networks (with 97% precision, 99% recall) and Rogers as an ISP (with 86% precision, 87% recall when using balanced data) highlights its practical utility in real-world scenarios. The methodology presented is adaptable to evolving telecommunications landscapes and can be recalibrated as network technologies advance. This research contributes to improving connectivity decisions in an increasingly digital world by translating complex network performance data into actionable insights. Integrating machine learning with telecommunications data establishes a foundation for future innovations in wireless communication optimization, supporting consumer experience and industry development in this essential infrastructure sector.

Acknowledgement. This research is fully funded by the Ministry of Higher Education, the Fundamental Research Grant Scheme (FRGS) (FRGS/1/2022/ICT08/UM/02/1).

References

1. Ding, Y.: Retention strategy for existing users of mobile communications. In: Lecture Notes on Data Engineering and Communications Technologies, vol. 84, pp.310–317https://doi.org/10.1007/978-981-16-5857-0_39
2. Kassim, M., Hisam, Z., Ismail, M.N.: Campus quality of services analysis of mobile wireless communications network signal among providers in Malaysia. Int. J. Adv. Comput. Sci. Appl. **13**, 181–187 (2022). https://doi.org/10.14569/IJACSA.2022.0130921

3. Ma, R., Teo, K.H., Shinjo, S., Yamanaka, K., Asbeck, P.M.: A GaN PA for 4G LTE-advanced and 5G: meeting the telecommunication needs of various vertical sectors including automobiles, robotics, health care, factory automation, agriculture, education, and more. IEEE Microw. Mag. **18**, 77–85 (2017). https://doi.org/10.1109/MMM.2017.2738498
4. Tafesse, W.: Social networking sites use and college students' academic performance: testing for an inverted U-shaped relationship using automated mobile app usage data. Int. J. Educ. Technol. High. Educ. **19**, 1–17 (2022). https://doi.org/10.1186/S41239-022-00322-0/FIGURES/1
5. Ghafar, A.A.; Kassim, M.; Ya'acob, N., Mohamad, R., Rahman, R.A.: Qos of Wi-Fi performance Based on signal strength and channel for indoor campus network. Bull. Electr. Eng. Inform. 9, 2097–2108 (2020). https://doi.org/10.11591/EEI.V9I5.2251
6. Pandey, S., Kadambi, G.R.: Modeling wireless mesh networks for load management. Int. J. Adv. Comput. Sci. Appl. **13**, 241–251 (2022). https://doi.org/10.14569/IJACSA.2022.0130530
7. Goldfarb, A., Tucker, C.: Digital economics. J Econ Lit **57**, 3–43 (2019). https://doi.org/10.1257/JEL.20171452
8. Whitacre, B., Gallardo, R., Strover, S.: Does rural broadband impact jobs and income? Evidence from spatial and first-differenced regressions. Ann. Reg. Sci. **53**, 649–670 (2014). https://doi.org/10.1007/S00168-014-0637-X
9. Rahul, Gupta, A., Raj, A., Arora, M.: IP traffic classification of 4G network using machine learning techniques. In: Proceedings - 5th International Conference on Computing Methodologies and Communication, ICCMC 2021, pp. 127–132 (2021). https://doi.org/10.1109/ICCMC51019.2021.9418397
10. Gebrie, H., Farooq, H., Imran, A.: What machine learning predictor performs best for mobility prediction in cellular networks? In: 2019 IEEE International Conference on Communications Workshops, ICC Workshops 2019 - Proceedings (2019). https://doi.org/10.1109/ICCW.2019.8756972
11. Riihijarvi, J., Mahonen, P.: Machine learning for performance prediction in mobile cellular networks. IEEE Comput. Intell. Mag. **13**, 51–60 (2018). https://doi.org/10.1109/MCI.2017.2773824
12. Elmasry, M.: Predict Network, Application Performance Using Machine Learning and Predictive Analytics. Theses (2019)
13. Eshti, S.A.: Drutology: a new system to measure customer's internet network experience in Bangladesh (2021)
14. Adeniyi, D.A., Wei, Z., Yongquan, Y.: Automated web usage data mining and recommendation system using K-nearest neighbor (KNN) classification method. Appl. Comput. Inform. **12**, 90–108 (2016). https://doi.org/10.1016/J.ACI.2014.10.001
15. Utamima, A., Aditya, F.I., Ashaari, F.W.G., Nugraha, A.K.: Applying machine learning algorithms for opinion mining on a digital internet service provider. In: ISDFS 2023 - 11th International Symposium on Digital Forensics and Security (2023). https://doi.org/10.1109/ISDFS58141.2023.10131886
16. Prajapati, C.D., Patel, A.K., Bhavsar, K.J.: Impact and challenges of data mining : a comprehensive analysis. Int. J. Sci. Res. Comput. Sci. Eng. Inf. Technol. **10**, 150–157 (2024). https://doi.org/10.32628/CSEIT241049
17. OSEMN Data Science Life Cycle - Data Science PM. https://www.datascience-pm.com/osemn/. Accessed 16 Jan 2025
18. Janc, K., Ilnicki, D., Jurkowski, W.: Spatial regularities in internet performance at a local scale: the case of Poland. Moravian Geog. Rep. **30**, 163–178 (2022). https://doi.org/10.2478/MGR-2022-0011

Bandwidth Burning: A Mechanism Against Whitewashing Attacks in BitTorrent-Like p2p File-Sharing Systems

John Ultra[1,2(✉)] and Susan Pancho-Festin[1]

[1] Department of Computer Science, UP Diliman, 1101 Diliman, Quezon City, Philippines
{jdultra,spfestin}@up.edu.ph

[2] Division of Natural Sciences and Mathematics, UP Tacloban College, Magsaysay Blvd, 6500 Tacloban City, Leyte, Philippines

Abstract. We present *bandwidth burning*, a form of resource burning, as a mechanism against whitewashing in BitTorrent-like peer-to-peer (p2p) file-sharing systems. Bandwidth burning imposed additional costs on attackers without putting a significant overhead on honest users. Under our proposed game-theoretic model, we showed that whitewashing is a dominated strategy. Numerical and simulation results showed the effectiveness of our proposed approach.

Keywords: free riding · whitewashing · BitTorrent · bandwidth burning

1 Introduction

Cheap or replaceable pseudonyms [1] provide anonymity and minimize the cost of joining distributed peer-to-peer (p2p) systems. However, this enables dishonest nodes to bypass the banning mechanism of p2p systems by leaving and rejoining with a new identity, called whitewashing attack [2].

To encourage user participation, it is common to see p2p systems like file-sharing networks provide unconditional or free benefits to new users. For example, in BitTorrent, leechers use optimistic unchoke to encourage other nodes to upload to them. Optimistic unchoke represents a node's willingness to cooperate even if reciprocation is not guaranteed. Self-interested users can exploit this mechanism in combination with whitewashing attacks for free-riding.

The problems that result from using cheap pseudonyms are easily avoided by so-called irreplaceable pseudonyms [1]. Irreplaceable pseudonyms can be implemented by public key infrastructure (PKI) in the form of an authenticated public key certificate at the individual or device level, or by some institutions whose business model relies on the validity of an individual identity such as credit card companies or financial institutions. Such services come with a cost and have implications for personal privacy. These make it non-viable in p2p systems designed to be public and whose users value their privacy.

Generally, whitewashing attacks can be discouraged by increasing the cost of generating new pseudonyms. The cost is either a monetary payment or a penalty in the form

L. Meng (Ed.): CCCN 2025, SCT, pp. 180–191, 2025.
https://doi.org/10.1007/978-3-032-03632-2_14

of differentiated services imposed on new users. This cost must be balanced to maintain continuous activity or participation within the system. It must not discourage genuine new users from joining. The idea is to encourage existing users to maintain their current identity by ensuring that optimal utility can only be achieved through continuous engagement with others.

In this paper, we revisit the problem of whitewashing in BitTorrent-like p2p systems. We proposed a practical mechanism called bandwidth burning to discourage users from whitewashing. The mechanism does not impose additional costs on honest users while increasing the cost of whitewashing on attackers.

2 Review of Literature

Several theoretical, experimental, and practical works have tackled the problem of whitewashing in different p2p settings. The problem of whitewashing and its relationship to free-riding was first investigated in [1] where the authors argued that social cost is necessary to prevent the collapse of social systems using cheap pseudonyms. For p2p systems, [2] presented a game-theoretic framework to analyze the effects of free-riding and whitewashing on the system performance of peer-to-peer (p2p) systems. Their results showed that exclusion policies are important, especially when societal generosity or contribution is low to prevent system collapse. They suggested the use of entry fees to deter whitewashing.

A similar line of work was pursued in [3–6]. [3] presented a game-theoretic analysis of the dynamics of (probabilistic) Tit-for-tat (TFT) against whitewashing in repeated games under the standard prisoner's dilemma payoff model. Similar to [2, 4] proposed and analyzed the effectiveness of a probabilistic stranger policy against whitewashing for a peer-to-peer (p2p) storage system. [5] proposed identity premium as a mechanism to discourage whitewashing in an online marketplace that uses reverse auction. Instead of requiring an entry fee, the authors proposed to penalize new users by requiring them to initially provide their services at a price lower than their winning bid price. In [6], the authors analyzed the dynamics of online communities as a repeated gift-giving game where the provider incurs a fixed cost c every time it fulfills the request of a client who obtains some benefit $b, b > c$.

In a more practical setting, [7] proposed a mechanism called Stranger Adaptive to deal with whitewashing. Using Stranger Adaptive policy, a node provides services to a stranger or new user with probability r denoting the recent generosity score of strangers or new users. r increases when the most recent stranger provided service and decreases if otherwise. Unlike approaches using static stranger policies [8], the penalty imposed by the mechanism depends on the level of free-riding in the system. However, when free-riding or whitewashing is high, new users are also penalized.

It is widely accepted that it is hard to distinguish a whitewasher from a genuine new user [1, 2]. [9] proposed a more general approach to address whitewashing. They called their approach observation preorder, which relies on the assumption that users are creatures of habit. This can be exploited by comparing the observed actions of new users with the past behaviors of banned users to identify potential whitewashers.

For mobile ad-hoc networks (MANET), [10] proposed a pay-before-use mechanism for new users. New users need to increase their reputation, e.g. by routing packets, to a minimum threshold to get services from the network.

The problem of whitewashing for BitTorrent-like peer-to-peer (p2p) file- sharing systems have been considered in [11–14]. [13] presented a BitTorrent- like peer-to-peer (p2p) file-sharing protocol that proposed using waiting time to deter whitewashing. New users are forced to wait for some time before they can download from other peers. This is implemented in the form of digitally signed messages generated by a tracker. Using game-theoretic analysis, they showed that whitewashing is a dominated strategy under a pay-off function that minimizes bandwidth consumption and download time. [11, 12, 14] applied trick-before-treat principle to address free-riding in peer-to-peer (p2p) file sharing systems. These mechanisms use cryptography to limit the free benefits obtained by free-riders. In turn, they can also limit whitewashing attacks. All of these approaches used game-theoretic analysis to justify the security of their constructions under the assumption that rational agents want to minimize bandwidth consumption.

[15] proposed Bazaar, a reputation system for online marketplace. Bazaar uses a risk network whose links have weights based on the accumulated monetary values of transactions between two users. The system prevents whitewashing by requiring new users to establish temporary links with existing users (i.e. a user vouches for the new user and is penalized if the new user misbehaves) or pay a deposit in an escrow amount. [16] also proposed using social networks to prevent whitewashing. More recently, [17] proposed BEQA, a trust and reputation management (TRM) that uses blockchain technology. User feedback is processed as blockchain transactions which by design require transaction fees to be finalized on the blockchain. This increases the cost of Sybil and whitewashing attacks.

3 Preliminaries

3.1 The File Sharing Model

Consider a vanilla BitTorrent-like p2p file-sharing with peers divided into two groups of leechers and seeders where a file f is divided into m equally sized pieces [18]. Let F be the set of pieces of file f. In the following discussion, we will refer to F and f interchangeably.

A new user joins the swarm by connecting to a tracker. Seeders unchoke leechers in a round-robin fashion. Seeders upload up to k free pieces to unchoked leechers in each period. For leechers, a Tit-for-tat (TFT) reciprocation policy is enforced when uploading to other leechers. To explore other potentially high-bandwidth peers in its peer set, a peer will occasionally unchoke a stranger peer, add it to its active set, and unconditionally upload k free pieces. At the start of a new round, a leecher bans any peer in its active set who did not meet the TFT policy in the previous round. We also assume that mutually trusted peers have a shared history [7], by sharing their local ban list with one another. Seeders can also use the shared history to identify selfish users and ban them from the swarm (i.e. indirect TFT)[1]. We also assume that a node's file status, the pieces completed by a node, is reliably shared among honest users[2].

Further, we assume that the system uses cheap and replaceable pseudonyms [1] so a banned user can rejoin using a new identity, i.e. a whitewashing attack. However, a user cannot use a Sybil attack i.e. an attacker can control only a single node (leecher) at a time.

3.2 Game-Theoretic Model

Game theory has been used to analyze the security of multi-party protocols under the notion of incentive-driven rational participants (for example, see [13]). In this subsection, we present the game-theoretic model that will be used in succeeding discussions.

Users and their Utility Functions. Let N be the set of leechers. Let $V_i(F)$ denote the value obtained by node $i \in N$ upon downloading the file F. Then i's utility $U_i = V_i(F) - C_i$ where C_i is a linear function denoting i's cost in downloading the file i.e., U_i is a quasi-linear function. We assumed that when i completes the file F, U_i is always positive. A user will never download a file that it does not value (*individual rationality*).

In the literature, the cost to a user in p2p file sharing is typically defined in terms of contributed upload bandwidth UB in bytes, consumed download bandwidth DB in bytes, or time spent downloading the file DT. Let us call these factors cost variables. Utility functions differ depending on the user's cost model. We present the following variations of possible leecher utility functions in terms of bandwidth and download time. Each function C defines the cost for a specific cost variable and type of user. We assume that download bandwidth is as valuable as upload bandwidth to a user and bandwidth cost is additive, i.e., $C^{BW}(UB) + C^{BW}(UB) = C^{BW}(UB + DB)$.

$$U_i^N = V_i(F) - \left[C^{DT}(DT_i) + C^{BW}(UB_i) + C^{BW}(DB_i)\right]$$

The first version of the utility function U^N considers both the download time and the bandwidth in the cost. We use this to represent the utility function of so-called honest-but-curious rational leechers. Honest-but-curious leechers would want to employ available strategies that would minimize both their bandwidth consumption and download time.

$$U_i^D = V_i(F) - [C_i^{BW}(UB_i) + C_i^{BW}(DB_i)]$$

The second utility function, U^D, represents users whose main priority is to minimize their bandwidth consumption. This group of users is typically called selfish leechers. However, unlike in the literature, we also define selfish users to be download time indifferent. Download time is not a factor to them. Since the download bandwidth cost should be essentially the same for types of users, selfish leechers can only attain maximum utility when cost $C^{BW}(UB_i) = 0$, i.e. $UB_i = 0$.

---

[1] We assume a reputation mechanism that can be used by seeders to gauge the quality of leechers.

[2] This information can be propagated through the network using the same mechanism used for disseminating the shared history.

In all cases, the only way for a user i to increase its utility is to minimize C_i, where C_i is the total cost incurred by i.

Finally, we have the following utility function for seeders. Let UB_i be the upload bandwidth contribution of leecher i.

$$U^S = V_S(\sum_{i \in N} UB_i)$$

We consider seeders altruistic, and bandwidth-cost indifferent. However, they want to maximize the total upload bandwidth contribution from leechers. We assume that the seeder's upload capacity is strictly lower than the swarm's combined download bandwidth requirements (i.e. seeder is a bottleneck).

3.3 Nash Equilibrium Strategy for Selfish Users

In the file-sharing model described above, an attacker can repeatedly perform whitewashing attacks, i.e., a *pure* whitewashing attack strategy, $\lceil m/k \rceil$ times to download the complete file F from a single leecher[3]. Specifically, this attack is a dominant strategy for a selfish user d because it can download the file without spending any upload bandwidth, $C^{BW}(UB_d) = 0$ and utilizing only the necessary download bandwidth, $DB_d = |F|$, to download all the pieces of F. Here, $|\cdot|$ denotes the size of f in bytes. Any other strategy where $C^{BW}(UB_d) > 0$ is dominated by the attack. In additional, under our individual rationality assumption, since d was able to complete the file, $(V_d(|F|) - C^{BW}(DB_d)) > 0$.

The same argument is not directly applicable for honest-but-curious users. For such users, the cost reduction from not using upload bandwidth may not be enough to outweigh the cost of longer download times, which can happen if all users decide to download only from the seeder who has limited upload bandwidth.

In the following section, we describe our proposed mechanism that will make whitewashing attacks a dominated strategy.

4 Proposed Mechanism

Our proposed approach, which we call *bandwidth burning*, uses randomization to increase the bandwidth cost incurred by dishonest users, i.e., selfish users, who use whitewashing attacks. The main idea is as follows:

- When a seeder unchokes (or a leecher optimistically unchokes) a new node i, it ignores i's pending piece requests. Instead, the k pieces it uploads to node i are randomly selected by the uploader based on i's file status.

In our proposed approach, we take away the ability of a new node to control the pieces it will receive from the seeder (leecher) which is the default behavior in BitTorrent. If a dishonest user performs a whitewashing attack, it rejoins the swarm with a new ID and a clean slate, i.e. its file status is empty[4]. Therefore, there is a significant chance that it will

[3] The shared history enables mutually trusting leechers to learn about dishonest users.

[4] Under our assumption that a node's file status is reliably shared among honest nodes.

receive duplicate pieces. Since the mechanism forces an attacker to waste resources like in proof-of-work (PoW), it can be considered a form of resource burning [19]. Selfish users will have to spend more download bandwidth with a strategy of pure whitewashing attacks, which makes it a dominated strategy under our model. We discuss this in detail in the following sub-sections.

4.1 Expected Amount of Download Bandwidth Consumption

Bandwidth consumption is one of the cost variables in leechers' utility functions. We can compute the average number of pieces that a selfish user, who uses whitewashing attacks, needs to download to obtain the complete set of distinct pieces under our proposed approach.

This problem can be modeled as an instance of a coupon collector's problem. One can imagine boxes of cereals, each containing a randomly chosen coupon. The problem translates to determining the number of boxes one must purchase to obtain a full collection. Informally, the problem asks to determine the number of boxes X_n^k one needs to purchase to complete a collection of n types [20] given each box contains k distinct coupons. The coupons are drawn randomly with replacement, each with a certain probability.

In our case, n corresponds to the number of distinct pieces of file F while k, the number of free pieces given to a new user, corresponds to the number of distinct coupons each box contains. For the basic case where each box contains one coupon, i.e., $k = 1$, and each coupon is drawn with equal probabilities, it is known that the average number of boxes or draws needed is:

$$E(X_n) = n(\frac{1}{n} + \frac{1}{n-1} + \frac{1}{n-2} + \cdots + \frac{1}{1})$$

The sum in the expression is the nth Harmonic number H_n. The nth Har- monic number H_n can be approximated by $ln(n) + \gamma$, where $\gamma \approx 0.57721$ is the Euler–Mascheroni constant, resulting to the following.

$$E(X_n) = nH_n \approx n(ln(n) + \gamma)$$

Notice that when $n > 4$, $H_n > 2$. We can apply this result to analyze our proposed approach when $k = 1$ i.e., a seeder gives exactly 1 free piece to a new user. As shown in Fig. 1, to collect all pieces in F, a selfish user using pure whitewashing attacks needs to download at least twice the size of F on average (indicated by the dotted black line). Under our assumption that the per byte cost of upload and download is the same, i.e. if $UB = DB$, then C^{BW} $(UB) = C^{BW}$ (DB), this makes the pure whitewashing attacks dominated by other strategies.

Let $BW_d = up_d + dw_{d,}$ be the total upload and download bandwidth consumed by user d to download file F where up_d and dw_d are the upload and download bandwidth consumed by d respectively. For ease of analysis, we assume that leechers obtain free pieces from the seeder and use a fair exchange mechanism to exchange pieces with one another. If d follows the protocol honestly, $up_d = n\text{-}k$ and $dw_d = n$ totaling to $BW_d = 2n\text{-}k$. Meanwhile, if d uses pure whitewashing attacks (against the seeder), $up_d = 0$

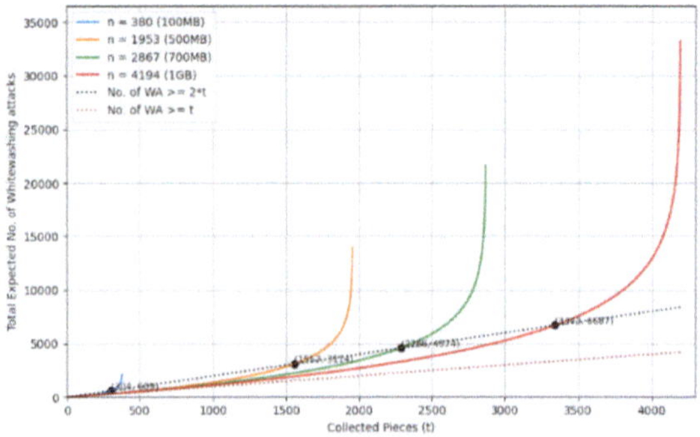

Fig. 1. The expected number of whitewashing attacks needed to obtain t distinct file pieces when the piece distribution is uniform. n denotes the number of pieces of a file with piece size $\approx 250KB$.

but $dw_d = E(X_n)$ where $E(X_n)$ is the expected number of WA needed to complete all n pieces. Since $E(X_n) > 2n\text{-}k$, the honest strategy dominates the pure whitewashing attack strategy. Since the pure whitewashing attack strategy is a dominated strategy, an attacker who wants to complete a file must switch to the honest strategy at some point. In doing so, it contributes its upload bandwidth thereby increasing the seeder's utility.

4.2 $E(X_n^k)$ When $k > 1$

We used numerical simulation to estimate $E(X_n^k)$ for $1 < k \leq 5$ under a uniform piece distribution. Figure 2 shows the inverse relationship between the $E(X_n^k)$ and k.

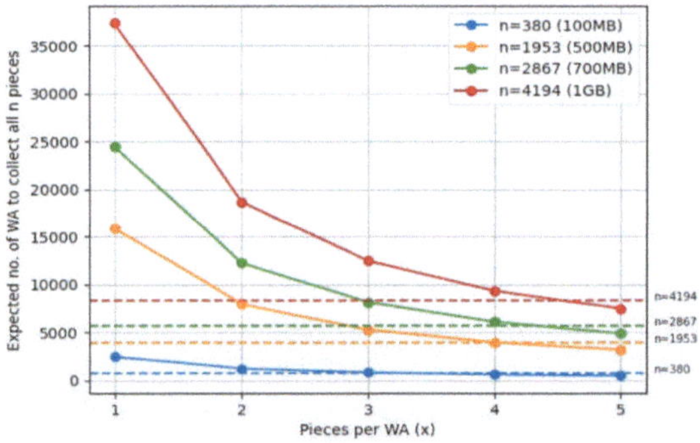

Fig. 2. Comparison of expected no. of whitewashing attacks for $1 \leq k \leq 5$ and varying file sizes. The dashed lines represent $y = 2n$ for different n.

$E(X_n^k)$ decreases as k increases specifically by a factor proportional to k. For a fixed k, $E(X_n^k)$ increases as n increases. This relationship is expressed in the formula below, which can be used to compute an upper bound of the ideal value of k relative to file size n.

$$E\left(X_n^k\right) \approx \frac{nH_n}{k} \tag{1}$$

$$2n < \frac{nH_n}{k} \tag{2}$$

$$k < \frac{H_n}{2} \tag{3}$$

5 Evaluation

Simulation experiments were conducted using the Peersim [21] P2P simulator. A partial implementation of BitTorrent in Peersim [22] was enhanced to implement BitTorrent's core protocol as well as the features described in Subsect. 3.1, and the bandwidth burning mechanism described in Sect. 4.

We evaluated the performance of the proposed mechanism by measuring the download time, share ratio, duplication factor, and piece completion rate of the different types of nodes under various parameters such as distribution of selfish users, piece selection mode, and whitewashing mode. For leechers, the share ratio is the ratio of the number of uploaded and downloaded pieces of a node. A value greater than 1 is ideal. The share ratio of seeders is the quotient of the number of uploaded pieces and the number of pieces in a file.

We define *duplication factor* as the ratio of the number of pieces downloaded by a node and a file's number of pieces. This captures the number of duplicated pieces downloaded by a node. *Piece completion rate* is the fraction of unique pieces downloaded by a node. A value of 1 means the node was able to complete the file. *Piece selection mode* is either vanilla for BitTorrent's default piece selection strategy or randomized for the proposed mechanism.

5.1 Simulation Setup

We used the following generic and BitTorrent-specific parameters in our experiments:

- *network size*- The initial number of nodes in the network. We set this to 100.
- *choke cycle time*- The length of the unchoke cycle used to control the number of pieces uploaded to a node. This takes the place of the variable k described previously. We set this to 10000 ms.
- *no. of seeders* - The number of initial seeders. We set this to 1 to capture
- the assumption that seeder bandwidth is a bottleneck.
- *file size*. The size of the file. We set this to 100MB.
- *piece size*. The size of each piece. We set this to 256 KB. A 100 MB file has 390 pieces.
- *no. of blocks*. The number of blocks per piece. We set this to 16.
- *node bandwidth*. The maximum upload and download bandwidth of a node. A node's bandwidth is randomly selected from the range: {640 Kbps, 1024 Kbps, 2048 Kbps, and 4096 Kbps}.

The experiments are run in a flash crowd setting where all nodes are added to the swarm at the beginning of the simulation. Honest leechers are set to leave the swarm once they complete their download. Other parameters were set on the script used to run the experiments. The experiments were run until the instantaneous average download rate of the swarm stayed close to 0 for at least 20 s. All experimental setups were repeated 30 times and results are presented with a 95% confidence interval. The experiments were carried out on a high-performance cluster with 8 machines each with 2 × AMD EPYC 7513 32-Core Processor and 16× 32 GB DRAM.

5.2 Results and Discussion

Figure 3a shows the piece completion rate of selfish users with the banning mechanism described in Sect. 3.1 with and without whitewashing. Without whitewashing, selfish users can only download at most 30% of the file's pieces. However, with whitewashing selfish users were able to download the file completely.

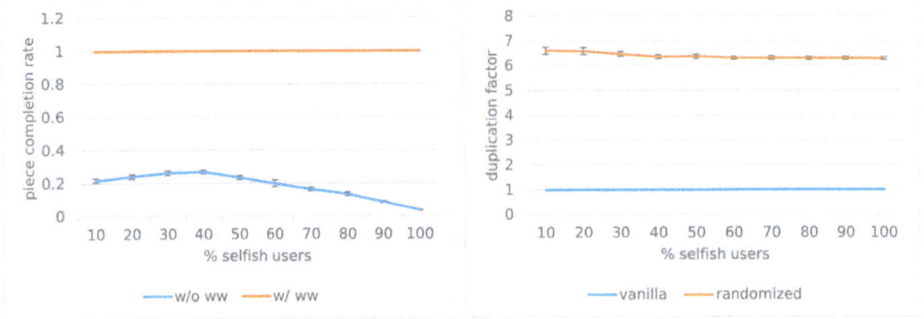

(a) piece completion rate (b) duplication factor

Fig. 3. The piece completion rate (with and without whitewashing) and duplication factor of selfish users.

Figure 3b shows the duplication factor of selfish users when downloading using whitewashing under vanilla (BitTorrent) and randomized (proposed) piece selection strategy. Though selfish users were able to download the file completely in both cases (as shown in Fig. 3a), they needed to download 6 times more than the file's number of pieces to complete the file under the proposed mechanism. This confirms the analytical results presented previously.

In addition, Fig. 4b shows that selfish users' average download time is at least one order of magnitude longer than the average download time of honest users. The difference in download time worsens as the number of honest users decreases.

Furthermore, Fig. 4a and Fig. 5a show that the proposed mechanism does not add a significant overhead on the download times and share ratio of honest users. Specifically, these results confirm that following the protocol is better than the whitewashing strategy in both download time and bandwidth consumption cost variables.

Finally, Fig. 3b showed the large duplication factor of selfish users under randomized piece selection while Fig. 5a showed that the proposed mechanism had no significant

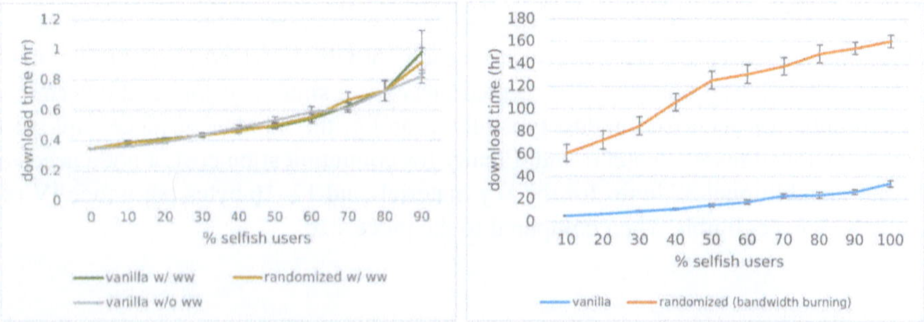

(a) honest (b) selfish

Fig. 4. The average download time of honest and selfish leechers.

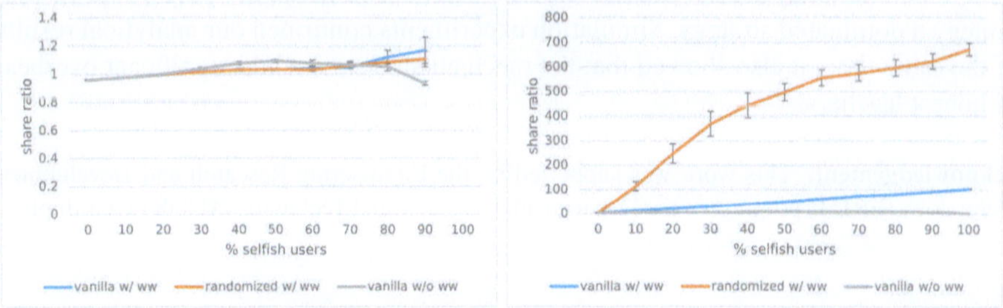

(a) leechers (b) seeder

Fig. 5. The average share ratio (SR) of honest leechers and the seeder.

effect on the share ratio of honest leechers (since they immediately leave the swarm once they finish downloading). As honest leechers complete their download, the seeder is left alone to cater to the requests of selfish users. This results in a significantly higher share ratio for the seeder in randomized piece selection than in vanilla piece selection as shown in Fig. 5b. However, this will only happen if selfish users become irrational and break the assumption about their utility function (see Sect. 3.2). In any case, this result is still consistent with our assumption that seeders are altruistic, and bandwidth cost indifferent.

Piece Encryption. In principle, if an attacker already has access to a subset $F\prime \subset F$, it can try to minimize the number of duplicate pieces downloaded by aborting the download of a piece (more precisely a piece's block) it already owns. This can be efficiently implemented using prefix-matching algorithms. We argue that this can be addressed by encrypting the pieces using a probabilistic encryption scheme. Like [14], the use of encryption does not add significant overhead due to the availability of fast implementation of secure symmetric encryption scheme like AES-GCM [23]. Our experiment showed that encrypting and encrypting a 256KB piece with AES-GCM takes about .23

ms on average[5]. For storage space, nodes only need to store an additional 256 bit of key material for each user and temporarily store each 256KB encrypted piece until uploading is completed. The number of encrypted pieces is a small constant and determined by the number of unchoked nodes, typically 4, and k, the number of pieces uploaded to each unchoked node in each round. Finally for communication cost, a node needs to transmit an additional 32 bytes for the key material, and 12–16 bytes per nonce/IV per piece which is negligible when compared to the piece size.

6 Conclusion

This paper explored analytically and experimentally bandwidth burning as a mechanism against whitewashing in BitTorrent-like peer-to-peer (p2p) file- sharing systems. Using game-theoretic analysis, we showed that the mechanism made the pure whitewashing strategy a dominated strategy. Simulation experiments confirmed our analytical results. At the same time, it also showed that the mechanism does not add significant overhead on honest leechers.

Acknowledgement. This work was supported by the Engineering Research and Development Technology (ERDT) program of the Department of Science and Technology (DOST), Philippines.

References

1. Friedman, E.J., Resnick, P.: The social cost of cheap pseudonyms. J. Econ. Manage. Strategy **10**(2), 173–199 (2001)
2. Feldman, M., Papadimitriou, C., Chuang, J., Stoica, I.: Free-riding and whitewashing in peer-to-peer systems. In: Proceedings of the ACM SIG- COMM Workshop on Practice and Theory of Incentives in Networked Systems, pp. 228–236 (2004)
3. Ouyang, J., Lin, Y., Zhou, S.: Game theoretic analyses for free-riding and whitewashing in p2p systems. In: Seventh International Symposium on Instrumentation and Control Technology: Measurement Theory and Systems and Aeronautical Equipment, vol. 7128, pp. 192–197 (2008). SPIE
4. Oualha, N., Roudier, Y.: A game theoretical approach in securing p2p storage against whitewashers. In: 2009 18th IEEE International Workshops on Enabling Technologies: Infrastructures for Collaborative Enterprises, pp. 128–133 (2009). IEEE
5. Vu, L.-H., Zhang, J., Aberer, K.: Using identity premium for honesty enforcement and whitewashing prevention. Comput. Intell. **30**(4), 771–797 (2014)
6. Zhang, Y., Park, J., Van Der Schaar, M.: Rating protocols in online com- munities. ACM Trans. Econ. Comput. (TEAC) **2**(1), 1–34 (2014)
7. Feldman, M., Lai, K., Stoica, I., Chuang, J.: Robust incentive techniques for peer-to-peer networks. In: Proceedings of the 5th ACM Conference on Electronic Commerce, pp. 102–111 (2004)

[5] The experiment involves running AES-GCM's encryption and decryption 10000 times with a 256 KB randomly generated plaintext. It was performed on a laptop with AMD Ryzen 5 4600H with 6 physical cores and 12 logical cores using Python's cryptography v41.0.7 package with OpenSSL backend.

8. Li, Y., Gruenbacher, D., Scoglio, C.: Evaluating stranger policies in P2P file-sharing systems with reciprocity mechanisms. Comput. Netw. **56**(4), 1470–1485 (2012). https://doi.org/10.1016/j.comnet.2011.12.017. Accessed 03 Sept 2023
9. Chen, J., Lu, H., Bruda, S.D.: A solution for whitewashing in p2p systems based on observation preorder. In: 2009 International Conference on Net- works Security, Wireless Communications and Trusted Computing, vol. 2, pp. 547–550. IEEE (2009)
10. Abbas, S., Merabti, M., Llewellyn-Jones, D.: Deterring whitewashing attacks in reputation based schemes for mobile ad hoc networks. In: 2010 IFIP Wireless Days, pp. 1–6. IEEE (2010)
11. Wang, J., Shen, R., Ullrich, C., Luo, H., Niu, C.: Resisting free-riding behavior in bittorrent. Futur. Gener. Comput. Syst. **26**(8), 1285–1299 (2010)
12. Berenjian, S., Hajizadeh, S., Atani, R.E.: An incentive security model to provide fairness for peer-to-peer networks. In: 2019 IEEE Conference on Application, Information and Network Security (AINS), pp. 71–76. IEEE (2019)
13. Backes, M., Ciobotaru, O., Krohmer, A.: RatFish: A file sharing proto- col provably secure against rational users. In: Hutchison, D., et al (eds.) Computer Security – ESORICS 2010. LNCS, vol. 6345, pp. 607–625. Springer, Heidelberg (2010). https://doi.org/10.1007/978-3-642-15497-3_37. Accessed 31 May 2022
14. Shin, K., Joe-Wong, C., Ha, S., Yi, Y., Rhee, I., Reeves, D.S.: T- chain: a general incentive scheme for cooperative computing. IEEE/ACM Trans. Netw. **25**(4), 2122–2137 (2017)
15. Post, A., Shah, V., Mislove, A.: Bazaar: strengthening user reputations in online marketplaces. In: 8th USENIX Symposium on Networked Systems Design and Implementation (NSDI 11) (2011)
16. de Pinninck, A.P., Schorlemmer, M., Sierra, C., Cranefield, S.: A social- network defence against whitewashing. In: Proceedings of the 9th Inter- national Conference on Autonomous Agents and Multiagent Systems: Volume 1-Volume 1, pp. 1563–1564 (2010)
17. Zeynalvand, L., Luo, T., Andrejczuk, E., Niyato, D., Teo, S.G., Zhang, J.: A blockchain-enabled quantitative approach to trust and reputation management with sparse evidence. In: Proceedings of the 20th International Conference on Autonomous Agents and MultiAgent Systems, pp. 1707–1708 (2021)
18. Cohen, B.: Incentives build robustness in bittorrent. In: Workshop on Economics of Peer-to-Peer Systems. Berkeley, vol. 6, pp. 68–72 (2003)
19. Gupta, D., Saia, J., Young, M.: Proof of work without all the work. In: Proceedings of the 19th International Conference on Distributed Computing and Networking, pp. 1–10 (2018)
20. Boneh, A., Hofri, M.: The coupon-collector problem revisited—a survey of engineering problems and computational methods. Stoch. Model. **13**(1), 39–66 (1997)
21. Montresor, A., Jelasity, M.: PeerSim: a scalable P2P simulator. In: Proceedings of the 9th International Conference on Peer-to-Peer (P2P'09), Seattle, WA, pp. 99–100 (2009)
22. Frioli, F., Pedrolli, M.: BitTorrent Protocol Peersim Simulator. https://peersim.sourceforge.net/code/bittorrent.tar.gz
23. Salowey, J., Choudhury, A., McGrew, D.: AES Galois Counter Mode (GCM) cipher suites for TLS (No. rfc5288) (2008)

Author Index

L. Meng (Ed.): CCCN 2025, SCT, pp. 193–194, 2025.
https://doi.org/10.1007/978-3-032-03632-2

MIX
Papier aus verantwortungsvollen Quellen
Paper from responsible sources
FSC® C105338

If you have any concerns about our products,
you can contact us on
ProductSafety@springernature.com

In case Publisher is established outside the EU,
the EU authorized representative is:
Springer Nature Customer Service Center GmbH
Europaplatz 3, 69115 Heidelberg, Germany

Printed by Libri Plureos GmbH
in Hamburg, Germany